Understanding Evolutionary Biology

Understanding Evolutionary Biology

Eden Cole

STATES
ACADEMIC PRESS
www.statesacademicpress.com

Published by States Academic Press,
109 South 5th Street,
Brooklyn, NY 11249, USA
www.statesacademicpress.com

Understanding Evolutionary Biology
Eden Cole

International Standard Book Number: 978-1-63989-537-3 (Hardback)

Cataloging-in-Publication Data

Understanding evolutionary biology / Eden Cole.
p. cm.
Includes bibliographical references and index.
ISBN 978-1-63989-537-3
1. Evolution (Biology). 2. Biology. 3. Evolution. I. Cole, Eden.

QH366.2 .U53 2022
575--dc23

TABLE OF CONTENTS

PREFACE

Biology is a branch of natural science which focuses on studying the physical structure and chemical processes related to living organisms. Evolutionary biology is concerned with the evolutionary processes which are responsible for the diverse life forms present on Earth. This includes a detailed investigation of common descent, natural selection and speciation. The discipline of evolutionary biology can be classified into theoretical biology, field biology, experimental biology, paleontology and evolutionary developmental biology. Some of the emerging areas of study under this discipline are sexual selection, genetic drift, mutation bias, molecular genetics, evolutionary synthesis, phylogenetics, etc. Evolutionary biology finds extensive applications in the fields of animal husbandry and drug development. This book attempts to understand the multiple branches that fall under the discipline of evolutionary biology and how such concepts have practical applications. It aims to shed light on some of the unexplored aspects of evolutionary biology. This book is an essential guide for both academicians and those who wish to pursue this discipline further.

A detailed account of the significant topics covered in this book is provided below:

Chapter 1- The biological process by which living organisms change their characteristics over several generations which sometimes leads to the development of a new species is known as evolution. This chapter explains evolution, microevolution and Darwin's theory in a detailed manner. This is an introductory chapter which will introduce briefly all significant aspects of evolution.

Chapter 2- The branch of biology which deals with the evolution and evolutionary processes which are responsible for diversity of life on Earth is known as evolutionary biology. This chapter discusses the various subfields of evolutionary biology like evolutionary developmental biology, evolutionary ecology, evolutionary genetics, and population genetics. The topics elaborated in this chapter will help in gaining a perspective about the subfields of evolutionary biology.

Chapter 3- The subfield of biology which includes the processes of growth and development of plants and animals is known as developmental biology. This chapter encompasses developmental biology, human development, plant development and animal development. All the diverse principles of developmental biology have been carefully analyzed in this chapter.

Chapter 4- The concept of evolutionary biology when two or more species have a common ancestor as an effect of speciation is known as common descent. Several other concepts of evolutionary biology like speciation, modern evolutionary synthesis, adaptation and genetic variation are discussed in this chapter. It elucidates the varied processes associated with this area of study.

Chapter 5- The biological process whereby organisms get adapted to the given environment in order to survive and reproduce is known as natural selection. This chapter contains topics like heritable traits, sexual selection, stabilizing selection, directional selection, negative selection, gene selection, etc. The aim of this chapter is to explore the field of natural selection.

Chapter 6- A change that occurs in any nucleotide of the genome of an organism either due to mistakes during replication or due to environmental factors is known as a mutation. This chapter discusses in detail the various aspects of mutations like mutationism, neutral mutations, point mutation, frameshift mutation, suppressor mutation, dynamic mutation and polar mutation.

It gives me an immense pleasure to thank our entire team for their efforts. Finally in the end, I would like to thank my family and colleagues who have been a great source of inspiration and support.

Eden Cole

Chapter 1

Introduction to Evolution

The biological process by which living organisms change their characteristics over several generations which sometimes leads to the development of a new species is known as evolution. This chapter explains evolution, microevolution and Darwin's theory in a detailed manner. This is an introductory chapter which will introduce briefly all significant aspects of evolution.

Evolution describes the process by which the diversity of life on earth developed over time from common ancestors. Within a population of organisms, there is variation in hereditary traits resulting from changes in the genetic code of individual organisms. These changes occur either by random mutation or by the shuffling of genetic material during sexual reproduction. One mechanism by which evolution, or change in populations occurs, is through natural selection. Evolution occurs with a population because individuals have different sets of these traits that affect their ability to survive and reproduce. For example, some individuals may be better at finding food, hiding from predators, or attracting mates. Those organisms with traits that help them to survive and reproduce are more likely to pass genes responsible for those traits to the next generation, a process termed natural selection. Over time, the distribution of characteristics in the population change, reflecting a change in the genetic makeup of the population. This process helps to account for the differences we see within species (microevolution) as well as the emergence of new species (macroevolution).

The scientific evidence for evolution is overwhelming. A classic example is the finch population on the Galapagos Islands. Galapagos finches have a variety of beak types adapted for cracking open different kinds of seeds. Changes in weather patterns have been shown to affect the kinds of seeds available—for example during drought, tough seeds that are difficult to crack are more abundant than soft seeds. Researchers found that finches with large beaks capable of cracking open those seeds were more likely to survive than finches with small beaks. As a result, these large beak finches became more abundant compared to finches with smaller beaks.

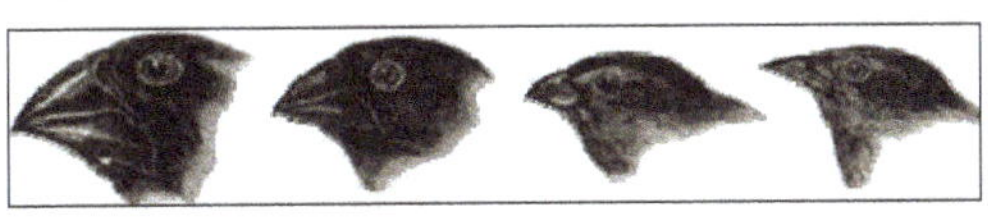

Geologists and paleontologists have documented systematic changes in organisms over hundreds of millions of years. By uncovering the fossil remains of once living organisms trapped in sedimentary rock layers, they find clear evidence that life forms have changed through time and that species have both emerged and disappeared. For example, mammalian fossils are evident in relatively recent rock layers, not in older layers; the oldest layers show no evidence of animals at all. Changes are evident within the human lineage as well, where the fossil record documents a number of intermediary forms between present day humans and now extinct non-human primates.

Evolution accounts for the similarities among diverse life forms. All species are remarkably alike at the cellular, molecular, and genetic levels, and all use DNA as the basic mechanism of heredity.

Even distantly related animals show similar patterns of development and striking anatomical similarities that are best accounted for by inferring a common ancestry. For example, even though their functions are different, there are clear structural similarities in the forelimbs of birds, frogs, and humans.

Evolution is not a phenomenon of the past. It is an active process occurring even now. The emergence of new strains of influenza, drug-resistant cancer cells, and pesticide-resistant insects demonstrate that the genetic makeup of populations changes over time by the process of natural selection. Within the population of influenza viruses, for example, some viruses are naturally resistant to the drugs used to treat them. As a result, the resistant viruses survive and reproduce and new influenza vaccines must be created to treat the newly evolved strain. Laboratory experiments also demonstrate evolution in action. Using rapidly reproducing species such as bacteria, yeast, and fruit flies, scientists have shown that altering the environmental conditions in which these organisms exist can induce genetic changes within the population.

Darwin's Theory

Darwinism is theory of the evolutionary mechanism propounded by Charles Darwin as an explanation of organic change. It denotes Darwin's specific view that evolution is driven mainly by natural selection.

Beginning in 1837, Darwin proceeded to work on the now well-understood concept that evolution is essentially brought about by the interplay of three principles: (1) variation—a liberalizing factor, which Darwin did not attempt to explain, present in all forms of life; (2) heredity—the conservative force that transmits similar organic form from one generation to another; and (3) the struggle for existence—which determines the variations that will confer advantages in a given environment, thus altering species through a selective reproductive rate.

On the basis of newer knowledge, neo-Darwinism has superseded the earlier concept and purged it of Darwin's lingering attachment to the Lamarckian theory of inheritance of acquired characters. Present knowledge of the mechanisms of inheritance is such that modern scientists can distinguish more satisfactorily than Darwin between non-inheritable bodily variation and variation of a genuinely inheritable kind.

Message

Darwin proposed a new explanation to account for the accepted phenomenon of evolution. He argued that the population of a given species at a given time includes individuals of varying characteristics. The population of the next generation will contain a higher frequency of those types that most successfully survive and reproduce under the existing environmental conditions. Thus, the frequencies of various types within the species will change over time.

There is an obvious similarity between the process of evolution as Darwin described it and the process by which the plant or animal breeder improves a domestic stock. The plant breeder selects the highest-yielding plants from the current population and (as far as possible) uses them as the parents of the next generation. If the characteristics causing the higher yield are heritable, then the next generation should produce a higher yield. It was no accident that Darwin chose the term natural selection to describe his model of evolution through differential rates of reproduction of different variants in the population. As a model for this evolutionary process, he had in mind the selection that breeders exercise on successive generations of domestic plants and animals.

We can summarize Darwin's theory of evolution through natural selection in three principles:

- Principle of variation: Among individuals within any population, there is variation in morphology, physiology, and behavior.
- Principle of heredity: Offspring resemble their parents more than they resemble unrelated individuals.
- Principle of selection: Some forms are more successful at surviving and reproducing than other forms in a given environment.

Clearly, a selective process can produce change in the population composition only if there are some variations among which to select. If all individuals are identical, no amount of differential reproduction of individuals can affect the composition of the population. Furthermore, the variation must be in some part heritable if differential reproduction is to alter the population's genetic composition. If large animals within a population have more offspring than do small ones but their offspring are no larger on average than those of small animals, then no change in population composition can occur from one generation to another. Finally, if all variant types leave, on average, the same number of offspring, then we can expect the population to remain unchanged.

Microevolution

Microevolution is defined as changes in allele frequency that can be observed within a population. In contrast to macroevolution, microevolution can be observed and measured in short periods of time, even within a single generation; macroevolution refers to the large-scale differences that can be observed between different species.

An accumulation of changes resulting from microevolution will eventually lead to macroevolution through the process of speciation. In this way, they are essentially the same process, although on a different time scale.

Processes of Microevolution

There are several ways in which microevolution can take place:

Gene Flow

Also known as "gene migration", gene flow is the transfer of alleles (variants of a gene) from one population to another. Gene flow occurs when individuals or their *gametes* are able to migrate between populations that are physically separated. For example, a person moving to a new country is adding to the gene pool of that country's population. Plants and fungi could have their pollen and spores distributed to another population by wind, pollinators or other mechanisms. In each gene flow event, the frequency of alleles present in each population is slightly altered.

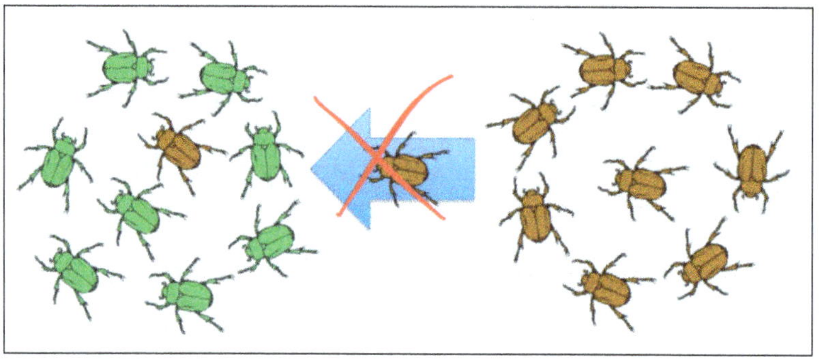

Usually gene flow is restricted to populations of the same species due to the barriers imposed by *reproductive isolation*. There are exceptions, however, such as the mechanisms employed by bacteria and viruses to transfer genes between organisms through *horizontal gene transfer*. In some cases, gene flow can occur between separate species, although generally only those that are closely related. If the two species have diverged recently, they may be able to produce offspring; however, these offspring are commonly infertile. Examples include the infertile 'mule' produced from a mating event between a donkey and a horse, or the 'liger' and 'tygon' resulting from the mating between a lion and a tiger.

Gene flow is a significant barrier to speciation; where gene flow is frequent between populations and the gene pools are regularly shuffled, the two populations do not have an opportunity to become genetically distinct.

Gene flow can also increase the genetic diversity of a population. For example, if an allele that has evolved within population B is introduced to population A, the genetic diversity of population A increases.

Genetic Drift

Genetic drift is the process of microevolution that describes the random fluctuations in the frequency of alleles within a population. Variants of genes (alleles) will become more or less numerous in a population if, by random chance, more copies of certain genes are passed on to the offspring generation. For example, if there are individuals with blue and red genes and, by chance, more individuals with blue genes reproduce, there will be more individuals with blue genes in the next generation. The frequency of blue genes may continue to increase in the following generations due to the chance of them being passed on to the offspring.

Genetic drift can lead to 100% fixation of an allele variant, or to it being lost entirely, both of which can result in a lower overall genetic diversity of the population. This is more likely in a small population and is common after *population bottlenecks*, in which case the rare alleles of the population can be lost rapidly.

Additionally, the *founder effect*, in which a small number of individuals becomes separated and isolated from a population, can cause genetic drift. The new population of individuals contains a limited sample of the full diversity of alleles in the original population, which will then increase in frequency.

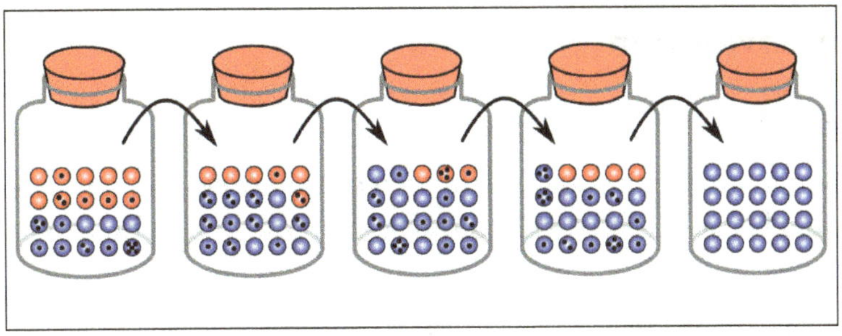

The image shows the probability chance of each gene being passed from parent to offspring through several generations.

In this case of microevolution, environmental or adaptive pressures do not determine the change in the frequency of allele variants in the population. The change is determined by random chance, and therefore the changes may be positive, negative or neutral for survival.

Mutation

Mutations are the ultimate cause of all variation in genetic diversity and the only source of new alleles. Although they occur rarely, a range of different circumstances, such as radiation, viruses and mutagenic chemicals, may cause them. They also occur naturally, as errors during DNA replication or meiosis.

A mutation may alter the outcome of the *trait* that the gene codes for, or it may prevent the gene from functioning correctly. However, mutations often have no effect or are only mildly deleterious (harmful), although they can also be advantageous. If the mutation is advantageous to the point where it gives the organism an adaptive advantage in an environment, it is likely to spread through the population through the process of *natural selection*, eventually leading to macroevolution.

The spread of mutated alleles can be even more rapid if they are *dominant alleles*, since their effects are shown immediately in the heterozygous and homozygous forms.

Selection

The *natural selection* of alleles that control for certain traits occurs when an organism's gene variants give it an adaptive advantage over the other gene variants in a population, ultimately allowing it to reproduce and increase the frequency of these genes.

Another way that alleles may be selected for is through *sexual selection*. In microevolution through sexual selection, organisms choose their mating partner based on an arbitrary feature that does not necessarily make them better adapted to their environment but that makes them more attractive to the opposite sex, for example, bright eyespots on a peacock's tail or large heavy antlers on a deer. Displaying the desired trait gives the mate a competitive advantage in reproduction, and eventually leads to an increase in the frequency of alleles associated with that trait.

Selection may also be controlled through *artificial selection* by humans. In this case, which is similar to sexual selection, certain desirable physical or behavioral traits of a plant or animal are chosen, and individuals that possess them are bred. The alleles become more common as a by-product of selecting for the favorable trait, even if it does not benefit the organism's survival.

Examples of Microevolution

Pesticide Resistance

Pesticides are designed to kill insects that might feed on a particular plant. Each time the pesticide is applied to the plant; a few individuals may survive, including those possessing certain genes that cause resistance to the pesticide.

The surviving individuals will go on to reproduce and increase the frequency of the genes that enabled them to survive in the population. Through subsequent applications of the pesticide, the number of individuals that have the genes for resistance will increase to complete fixation; at this point, the pesticide cannot kill any of the individuals, as they are all resistant. This means a new pesticide must be developed to kill future generations of insect. The mechanisms of microevolution occurring here are natural selection and genetic drift by means of a bottleneck.

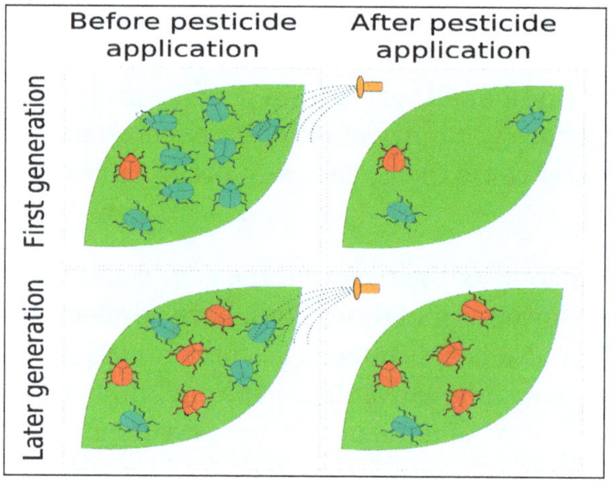

Evolution of Viral Disease

Although they are a highly diverse group, viruses comprise several prerequisites in their biology that render them particularly susceptible to rapid microevolution. Unlike other organisms, viruses may store their genetic information using either DNA or RNA. Mutation rates during replication are much higher in RNA than in DNA and therefore new genes, which may be beneficial to the organism and increase in frequency through selection, are created more frequently. Viruses also have large population sizes and short generation times, which further increases the probability of a possible genetic mutation and allows mutations to spread rapidly through populations.

Additionally, some viruses, such as influenza and HIV, can 'shuffle' their genes with other strains of the virus in a process called *recombination*. This can lead to the evolution of entirely new viral strains and greatly increases the genetic diversity of the population.

The constant emergence and microevolution of viruses due to their high mutation rate makes them incredibly hard to treat when they infect other animals, as the populations can quickly develop resistance to drugs.

Macroevolution

Macroevolution refers to the concept of large-scale evolution that occurs at the level of species and above. Macroevolution can be used to describe the differences between two closely related but distinct species, such as the Asian Elephant and the African Elephant, which cannot mate due to the barriers imposed by reproductive isolation. This is the process of speciation, which can be driven by a number of different mechanisms. Additionally, macroevolution can describe differences between those organisms belonging to larger clades of organisms, for example the different taxonomic groups within the primates.

The term macroevolution can also be used to explain the shared common ancestry between all living organisms, a concept known as *Universal Common Descent*. This describes the derivation of all existent and extinct life forms from a single origin, and includes evolutionary milestones such as the origins of plants, mammals, reptiles, birds, fish, non-avian dinosaurs and more.

The term 'macroevolution' is often used in contrast to the within-species genetic changes that relate to *microevolution*, although the two concepts are fundamentally the same, albeit on different time scales; each of the evolutionary mechanisms—*mutation*, *gene flow*, *genetic drift* and *natural selection*—that alter the *gene pool* of a population through microevolution, will accumulate over a long time period, resulting ultimately in macroevolution. In the case of Universal Common Decent, microevolution has been driving the macroevolution of living organisms for 3.8 billion years (that's 3,800,000,000 years).

While macroevolution can occur due to an accumulation of micro-evolutionary changes, a random event that somehow separates a population into two different populations may also be responsible for the formation of new species. This is called *allopatric speciation*. For example, individuals in a population of lizards that live in a valley may become separated by the flow of newly formed river. The river makes it impossible for the individuals on each side to interact and breed. Due to the climate and landscape, one side of the valley is lush green vegetation; the other is hot and arid. Over time, each lizard population will adapt to the conditions on each side of the valley. If the interactions between them continue to be restricted, the two populations could become so different that if they were able to reunite, they could no longer breed. The two populations of lizards would be classed as new species and would have increased to the world's biodiversity through macroevolution.

Creationists more commonly dispute the factual legitimacy of macroevolution than microevolution because microevolution is demonstrable within a lab, whereas macroevolution cannot be observed within the lifetime of a human. There are, however, many ways to observe macroevolution using the evidence available from fossils, geology and radiometric dating, genetics, and the ecology, morphology and behavior of living organisms.

Examples of Macroevolution

The Evolution of Sirenians

The sirenians are a group of mammals within the placental mammals, consisting of the manatees and the Dugong, as well as other extinct species. Also known as the 'sea cows', the sirenians feed solely on sea grass and are the only *herbivorous* aquatic mammals. They share a common ancestor with the elephants and the extinct mammoths, in a taxonomic group called the *Tethytheria*. The evidence for the macroevolution that has separated these groups of animals comes from *vestigial structures* within their bodies.

Firstly, the flippers of sirenians have internal bone structures that are *homologous* to the bone structure of all other terrestrial *tetrapods* (four legged animals). Each flipper contains an upper arm bone; two forearm bones, wrist bones, hand bones and five finger bones. On the external tips of their flippers the sirenians have toenails, which are the same as those seen in elephants.

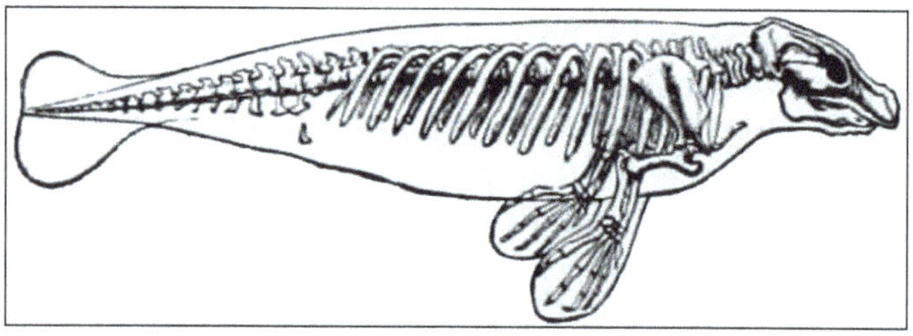

The image shows the skeleton of a manatee. The homologous bone structures are visible in the flippers. Like all other terrestrial mammals, manatees have five finger bones although they have no fingers. Furthermore, manatees don't have hind limbs but they do have pelvic bones, which are usually used to support the hind limbs in other tetrapods. Because they lack hind limbs they also don't have femurs but they have vestigial hip sockets where the femur attaches to the pelvis.

Solidifying the evidence that the sirenians share a common ancestor with other *Tethytheria*, and have evolved through macroevolution, a 50 million year old fossilized skeleton of a sirenian was found in Jamaica. The fossil had the features of a modern sirenian—including a solid boned skeleton to weigh it down in the water and nostrils on the back its head allowing it to breath at the surface—but it also had four well-developed legs! Several other similar fossil skeletons have also been found with the hind legs, although reduced in size; so reduced that they would not have been able to support the weight of the animal.

From the modern physical evidence of the vestigial structures, as well as the fossil evidence, it is clear that the sirenians share a common ancestor with modern elephants. Originally land-dwelling animals, the ancestors of the sirenians would have utilized the food source of sea grass within marine environments. As the hind limbs are not much use in the water, these became gradually smaller over millions of years until disappearing completely, leaving only some boney reminders of their land-based ancestors.

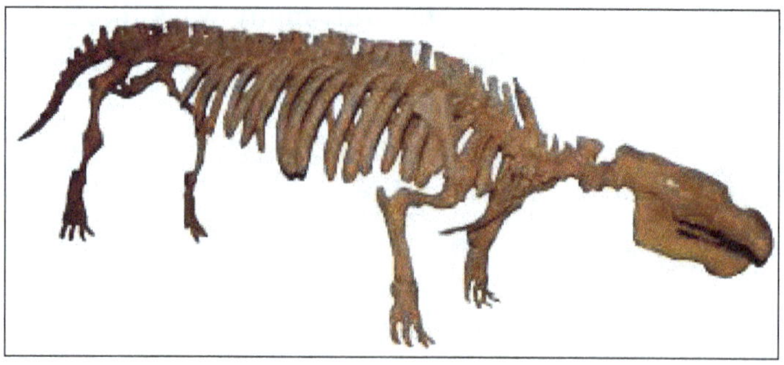

The image shows the skeleton of the Pezosiren portelli the ancestral form of the sirenians, which has fully formed legs and the other features of modern manatees and dugongs.

Hemoglobin Differences

It is possible to view the evidence for macroevolution through the similarities and differences between the genetic information stored within all living organisms. The genetic information, which determines the characteristics that will be displayed in any organism, is held within strands of *nucleotide bases* called DNA. The sequence of the nucleotides within the DNA is used as a code, giving precise instructions for the production of *amino acids*; amino acids are the 'building blocks' of proteins, which in turn build living cells.

A strong piece of evidence for the theory of Universal Common Descent comes from the fact that every living organism—except some *prokaryotes* that contain only RNA, which split from other lineages very early in evolutionary history—uses DNA consisting of the same nucleotides, arranged in the same manner, to code for the production of amino acids. Furthermore, the proteins of all organisms are comprised of combinations of exactly the same 20 amino acids, without exception.

Over time and during reproduction, the genetic information is replicated and slight mutations in the sequences of nucleotides occur, resulting in slightly altered DNA strands and thus altered sequences of amino acids and proteins.

Because of these mutations, it is possible to work out how related two organisms are to each other, based on the differences and similarities of their amino acids. This is well demonstrable with the sequence of amino acids present in a protein that is present in nearly all animals, hemoglobin. Used to transport oxygen around the body, hemoglobin is made up of two sets of identical amino acid chains: one pair consisting of 141 amino acids and the other consisting of 146.

When compared with the amino acid sequences of humans it is possible to see how far diverged we are from certain other organisms, for example the number of amino acid differences in hemoglobin of:

- Gorillas – 1
- Rhesus Monkeys – 8
- Mice – 27
- Chickens – 45
- Frogs – 67
- Lampreys – 125

The ancestors of lampreys split from the jawed fish around 500 million years ago, and the ancestors of humans split from gorillas around 10 million years ago. The differences and similarities between the amino acid sequences of hemoglobin can therefore support the common ancestry between species as well as providing evidence for the times of divergence.

Convergent Evolution

Convergent evolution is the process by which unrelated or distantly related organisms evolve similar body forms, coloration, organs, and adaptations. Natural selection can result in evolutionary convergence under several different circumstances. Species can converge in sympatry, as in mimicry complexes among insects, especially butterflies (coral snakes and their mimics constitute another well-known example). Mimicry evolves after one species; the 'model' has become aposematic (warningly colored) because it is toxic or poisonous and therefore protected. Two distinct kinds of mimicry are recognized, Batesian and Müllerian. In Batesian mimicry, the mimic is palatable or unprotected, but gains from being mistaken for the model, which is unpalatable or protected. Two protected model species can also converge because of the advantage of being mistaken for each other (Müllerian mimicry).

Left: Viceroy Butterfly (The mimi - palatable species).
Right: Monarch Butterfly (The model - distasteful species).

Mimicry is an interesting consequence of warning coloration that nicely demonstrates the power of natural selection. An organism that commonly occurs in a community along with a poisonous or distasteful species can benefit from a resemblance to the warningly colored species, even though the 'mimic' itself is nonpoisonous and/or quite palatable. Because predators that have experienced contacts with the model species, and have learned to avoid it, mistake the mimic species for the model and avoid it as well. Such false warning coloration is termed Batesian mimicry after its discoverer.

Many species of harmless snakes mimic poisonous snakes; in Central America, some harmless snakes are so similar to poisonous coral snakes that only an expert can distinguish the mimic from the 'model.' A few experts have even died as a result of a superficial misidentifications. Similarly, certain harmless flies and clearwing moths mimic bees and wasps, and palatable species of butterflies mimic distasteful species. Batesian mimicry is disadvantageous to the model species because some predators will encounter palatable or harmless mimics and thereby take longer to learn to avoid the model. The greater the proportion of mimics to models, the longer is the time required for predator learning and the greater the number of model casualties. In fact, if mimics became more abundant than models, predators might not learn to avoid the prey item at all but might actively search out model and mimic alike. For this reason Batesian mimics are usually much less abundant than their models; also, mimics of this sort are frequently polymorphic (often only females are mimics) and mimic several different model species.

Müllerian mimicry is different, and occurs when two species, either distasteful or dangerous, mimic one another. Both bees and wasps, for example, are usually banded with yellows and blacks. Because potential predators encounter several species of Müllerian mimics more frequently than just a single species, they learn to avoid them faster, and the relationship is actually beneficial to both prey species. The resemblance need not be as precise as it must be under Batesian mimicry because neither species actually deceives the predator; rather, each only reminds the predator of its dangerous or distasteful properties. Müllerian mimicry is beneficial to all parties including the predator; mimics can be equally common and are rarely polymorphic.

Molecules can evolve convergently, especially when parasites mimic molecular messages that signal 'self' to immune responses of hosts, which allow the parasite to elude its host's defenses. Molecular convergence could also take place when a particular metabolic function requires similar or identical molecular structure. Some gene circuits and gene networks appear to have undergone convergent evolution by single-gene duplications in higher eukaryotes. Convergence in DNA nucleotide sequences would lead to erroneous phylogenetic conclusions, which would be problematical for molecular systematic studies.

Evolutionary convergence involving unrelated organisms living in similar environments but in different places (allopatry) can also occur in another way. This usually takes place in relatively simple communities in which biotic interactions are highly predictable and the resulting numbers of different ways of exploiting the environment are limited. Similar environments pose similar challenges to survival and reproduction, and those traits that enhance Darwinian fitness are selected for in each environment. Such organisms that fill similar ecological roles in different, independently-evolved, biotas are termed "ecological equivalents". Examples are legion.

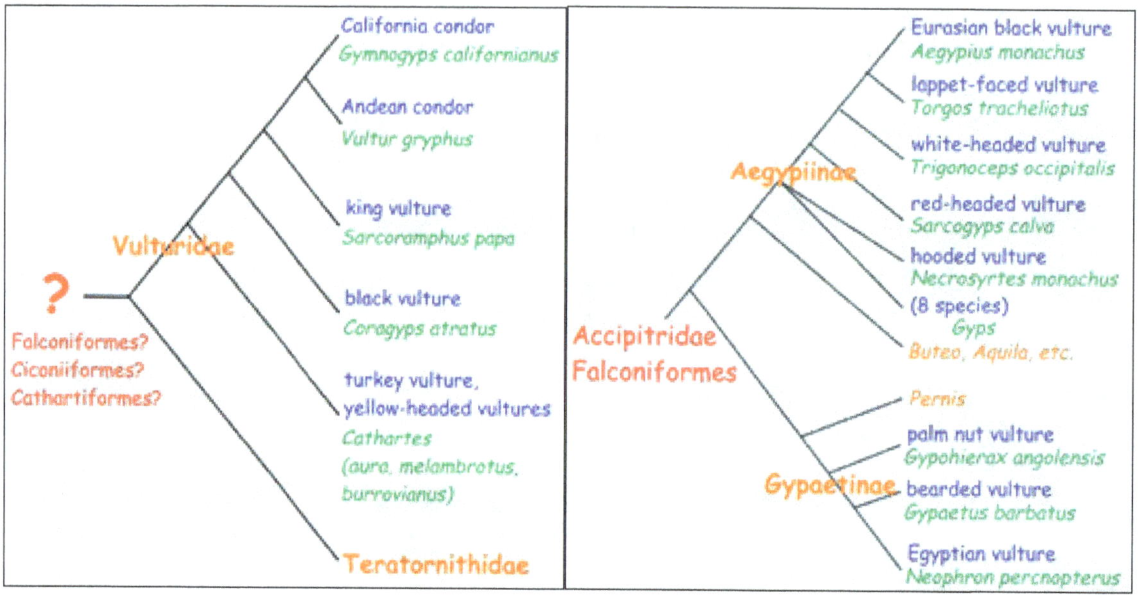

Wings and wing-like structures have evolved independently several times, in insects, reptiles (pterosaurs and birds) and in mammals (bats). Flight first evolved in insects about 330 million years ago (mya), second in pterosaurs (about 225 mya), later in birds (about 150 mya), and still later in bats (50-60 mya). Some frogs, lizards, and mammals have also evolved the ability to glide, presumably a precursor to flight. In order to land safely, such hang gliders must time their stall precisely at the right moment and place.

New World Cactus African Euphorb

For many years, avian systematists classified Old World and New World vultures as close relatives, both thought to be allied to raptors (hawks and owls). However, DNA hybridization suggested that, although Old World vultures are indeed related to raptors, New World vultures are not, but are descendents of common ancestors to storks and cranes (more recent studies are equivocal but still support independent evolution of the two clades). Interestingly, a behavioral trait was conserved in the evolution of new world vultures: when heat stressed, storks defecate/urinate on their own legs to dissipate excess heat.

Arid regions of South Africa support a wide variety of euphorbeaceous plants, some of which are strikingly close to American cacti phenotypically. They are leafless stem succulents, protected by sharp spines, presumably adaptations to reduce water loss and predation in arid environments. Similarly, evergreen sclerophyll woody shrubs have evolved convergently under Mediterranean climates in several different regions.

African Macronix American Medolark

A brown bird of some African prairies and grasslands, the African yellow-throated longclaw (*Macronix croceus*), a motacillid, has a yellow breast with a black chevron "V". This motacillid looks and acts so much like an American meadowlark (*Sturnella magna*), an icterid, that a competent bird watcher might mistake them for the same species, yet they belong to different avian families. Another example is the North American Little Auk and the Magellan Diving Petrel, two superficially very similar aquatic birds, which belong to different avian orders.

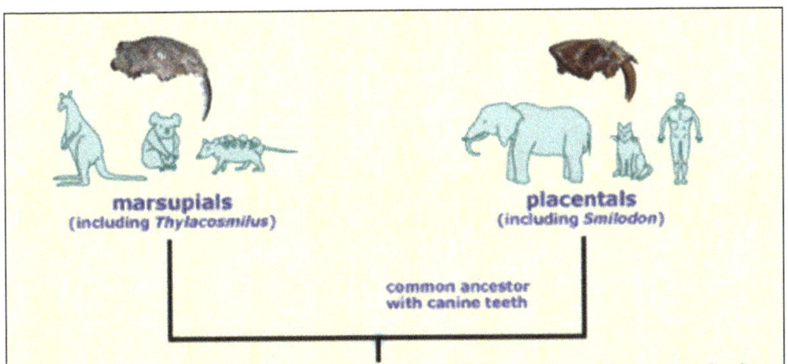

South American *Thylacosmilus* North American *Smilodon*

Convergence sometimes occurs under unusual conditions where selective forces for the achievement of a particular mode of existence are particularly strong. Presumably in response to thick-skinned prey, two fossil saber-tooth carnivores, the South American marsupial 'cat,' *Thylacosmilus*, and the North American placental saber-toothed tiger, *Smilodon*, evolved long knife-like canine teeth independently (but these were not contemporary). Many other marsupial mammals have undergone convergent evolution with placentals, including moles, mice, wombats (woodchucks), numbats (anteaters), quolls (cats), and thylacines (wolves).

Still another example of convergent evolution is seen in the similar shape and coloration of fish and cetaceans, both of which have adapted to the marine environment by developing a fusiform body and neutral buoyancy (an extinct group of marine reptiles known as ichthyosaurs evolved the same body plan). Sharks and dolphins are also counter shaded, with a light underbelly and a darker upper surface, which makes them less visible from both below and above. However, countershading is actually the rule among both arthropods and vertebrates, so it is presumably an ancestral state that has been retained throughout the evolution of both groups.

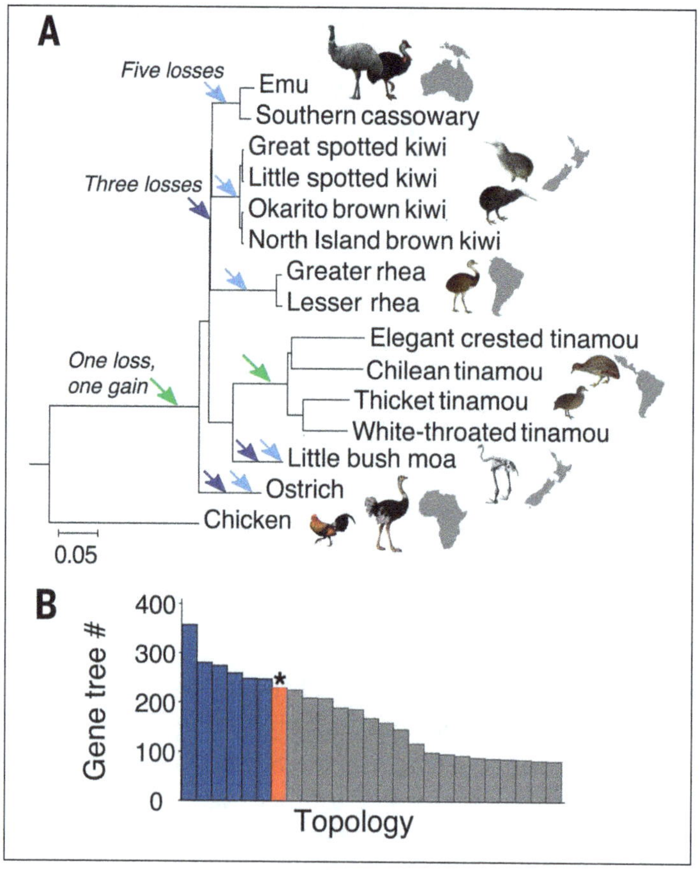

Flightless birds such as the emu, ostrich, and rhea fill very similar ecological niches on different continents. If ratites evolved from a Gondwanan common ancestor, they would not represent evolutionary convergence but instead would constitute an example of a shared (and conserved) ancestral flightless state. Now thought to be convergent, DNA evidence suggests that these "ratites" do not share a common ancestry but have evolved independently.

Live bearing, or viviparity, has evolved over 100 times among squamate reptiles (lizards and snakes), usually in response to cold climates. The probable mechanism behind the evolution of viviparity is that, by holding her eggs, a gravid female can both protect them from predators and, by basking, warm them, which would increase rate of development. Eventually, such selective forces favoring egg retention could lead to eggs hatching within a mother and live birth. This has happened even in geckos, all of which lay eggs except for one genus in New Caledonia and several related cold temperate New Zealand forms, which bear their young alive. In some skinks and xantusiid lizards, embryos attach to their mother's oviducts and grow, gaining nutrients during development via placental arrangements reminiscent of those in mammals.

Thorny Devil Moloch horridus.

Horned Lizard Phrynosoma cornutum.

Convergent evolutionary responses of lizards to arid environments are evident between continents. For example, Australian and North American deserts both support a cryptically-colored and thornily-armored ant specialized species: the thorny devil, *Moloch horridus*, an agamid, exploits this ecological role in Australia, while its counterpart the desert horned lizard (*Phrynosoma*), an iguanid, occupies it in North America. No Kalahari lizard has adopted such a life style. Interestingly, morphometric analysis demonstrates that the thorny devil and desert horned lizard are actually anatomically closer to one another than either species is to another member of its own lizard fauna, which are much more closely related.

Emerald Tree Boa Corallus caninus.

Green tree python Chondropython viridis.

Emerald Tree Boas from South American Amazonian rainforests are strikingly convergent with Green Tree Pythons found halfway around the world in similar rainforests in Australia, a spectacular example of ecological equivalents. Both of these snakes live high up in the canopy and eat birds. Adults are green, cryptically colored, matching the colors of leaves. Juveniles of both species are bright yellow or orange.

Juvenile Emerald Tree Boa.

Juvenile Green Tree Python.

Colorful, blue, red and yellow tails have evolved repeatedly among distantly related lizards in many families (agamids, anguids, gymnopthlamids, lacerids, skinks, and teiids), presumably a ploy to attract a predator's attention away from the head to the tail, which can be broken off and regenerated should a predator attack it.

Blue-tailed anguid from Mexico (Celestes).

The New World iguanid *Basiliscus*, sometimes called 'Jesus lizards,' because they can run across the surface of water, have undergone convergent evolution with the Old World agamid *Hydrosaurus*. Both *Basiliscus* and *Hydrosaurus* have enlarged rectangular, plate-like, fringed scales on their toes, which allow these big lizards to run across water using surface tension for support.

A skink (Morethia butleri) from Australia.

Open sandy deserts pose severe problems for their inhabitants: (1) windblown sands are always loose and provide little traction; (2) surface temperatures at midday rise to lethal levels; and (3) open sandy areas offer little food or shade or shelter for evading predators. Even so, natural selection over eons of time has enabled lizards to cope fairly well with such sandy desert conditions. Subterranean lizards simply bypass most problems by staying underground, and actually benefit from the loose sand since underground locomotion is facilitated. Burrowing is also made easier by evolution of a pointed, shovel-shaped head and a countersunk lower jaw, as well as by small appendages and muscular bodies and short tails. Such a reduced-limb adaptive suite associated with fossorial habits has evolved repeatedly among squamate reptiles in both lizards and snakes.

During the hours shortly after sunrise, but before sand temperatures climb too high, diurnal lizards scurry about above ground in sandy desert habitats. Sand specialized lizards provide one of the most striking examples of convergent evolution and ecological equivalence. Representatives of many different families of lizards scattered throughout the world's deserts have found a similar solution for getting better traction on loose sand: enlarged scales on their toes, or lamellae, have evolved independently in six different families of lizards: skinks, lacertids, iguanids, agamids, gerrhosaurids, and geckos. A skink, *Scincus*, appropriately dubbed the 'sand fish,' literally swims through sandy seas in search of insect food in the Sahara and other eastern deserts. These sandy desert regions also support lacertid lizards (*Acanthodactylus*) with fringed toes and shovel noses. Far away in the southern hemisphere, on the windblown dunes of the Namib desert of southwestern Africa, an independent lineage of lacertids, *Meroles*(formerly *Aporosaura) anchietae*, has evolved a similar life form. In North America, this body form has been adopted by members of the iguanid genus *Uma*, which usually forage by waiting in the open and eat a fairly diverse diet of various insects, such as sand roaches, beetle larvae and other burrowing arthropods. They also listen intently for insects moving buried in the sand, and dig them up. Sometimes they dash, dig, and paw through a patch of sand and then watch the disturbed area for movements.

Meroles (Aporosaura) anchiete.

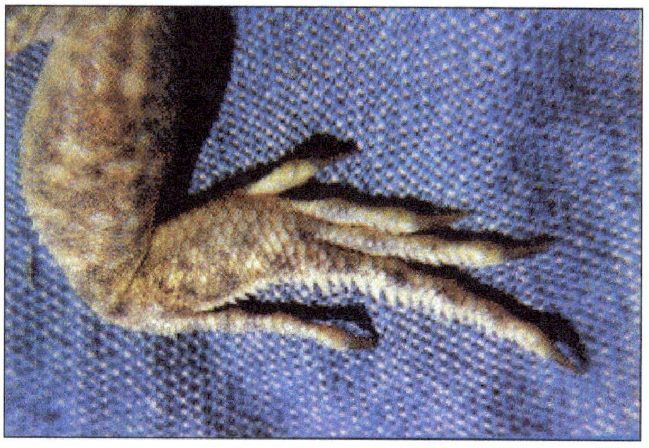

Toes of Uma scoparia.

All of these lizards have flattened, duckbill-like, shovel-nosed snouts, which enable them to make remarkable 'dives' into the sand even while running at full speed. The lizards then wriggle along under the surface, sometimes for over a meter. One must see such a sand diving act to appreciate fully its effectiveness as a disappearing act. Some Namib Desert lizards have discovered another solution to gain traction on powdery sands: frog-like webbing between the toes as seen in the geckos *Kaokogecko* and *Palmatogecko* (now *Pachydactylus*).

Pachydactylus (formerly Palmatogecko) rangeri.

Other lizards that have undergone convergent evolution include rock mimics such as the North

American horned lizard *Phrynosoma modestum*, an iguanid, and the Australian agamid *Tympanocryptis cephalus*. New World teiids (*Tupinambis*) have converged on Old World varanids (*Varanus*): members of both genera are large predatory lizards with forked tongues which they use as edge detectors to find scent trails and track down their prey (other vertebrates).

Rock Mimic Tympanocryptis cephalus.

Sometimes, roughly similar ecological systems support relatively few conspicuous ecological equivalents but instead are composed largely of distinctly different plant and animal types. For instance, although bird species diversities of temperate forests in eastern North America and eastern Australia are similar, many avian niches appear to be fundamentally different on the two continents. Honeyeaters and parrots are conspicuous in Australia whereas hummingbirds and woodpeckers are entirely absent. Apparently different combinations of the various avian ecological activities are possible; thus, an Australian honeyeater might combine aspects of the food and place niches exploited in North America by both warblers and hummingbirds. An analogy can be made by comparing the 'total avian niche space' to a deck of cards. This niche space can be exploited in a limited number of ways, and each bird population or species has its own ways of doing things, or its own "hand of cards," determined in part by what other species in the community are doing.

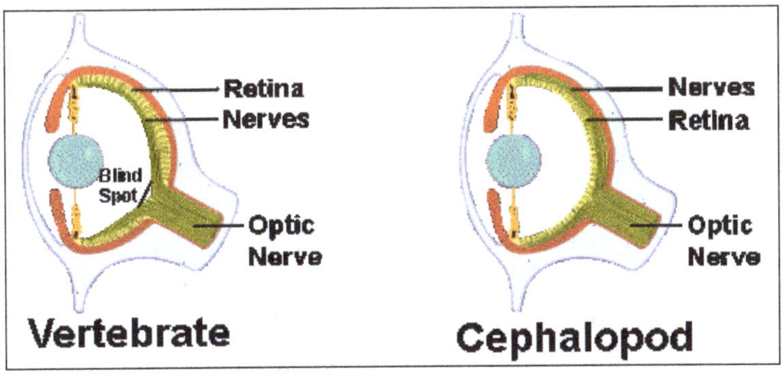

A very striking example of evolutionary convergence involves the eyes of vertebrates and cephalopod mollusks. Both have independently evolved complex camera-like eyes complete with an aperture, lens, and retina. Prominent anti-Darwinian Charles Hodges once suggested that the vertebrate eye was too complex to have evolved by natural selection and therefore must have been "designed." However, vertebrate eyes are poorly designed as compared to cephalopod eyes. In vertebrates, nerve fibers pass in front of the retina creating a blind spot, whereas nerves lie behind the retina in the superior cephalopod eye which does not have a blind spot.

Coevolution

Yucca moth (Tegeticula yuccasella): A female yucca moth (Tegeticula yuccasella) pushing pollen into the stigma tube of the yucca flower while visiting the flower to deposit her eggs.

Coevolution is the process of reciprocal evolutionary change that occurs between pairs of species or among groups of species as they interact with one another. The activity of each species that participates in the interaction applies selection pressure on the others. In a predator-prey interaction, for example, the emergence of faster prey may select against individuals in the predatory species who are unable to keep pace. Thus, only fast individuals or those with adaptations allowing them to capture prey using other means will pass their genes to the next generation. Coevolution is one of the primary methods by which biological communities are organized. It can lead to very specialized relationships between species, such as those between pollinator and plant, between predator and prey, and between parasite and host. It may also foster the evolution of new species in cases where individual populations of interacting species separate themselves from their greater metapopulations for long periods of time.

How an interaction coevolves between species depends not only on the current genetic makeup of the species involved but also on new mutations that arise, the population characteristics of each species, and the community context in which the interaction takes place. Under some ecological conditions (such as in some predator-prey interactions or between competitors for a resource), an antagonistic interaction between two species can coevolve to enhance the antagonism; the species "build up" methods of defense and attack, much like an evolutionary arms race. Under other ecological conditions (such as in certain parasite-host interactions), however, the antagonism may be lessened.

Coevolution does not necessarily require the presence of antagonism. The interactions or characteristics within groups of unrelated species may converge to allow individual species to exploit valuable resources or enjoy increased protection. Once an interaction evolves between two species, other species within the community may develop traits akin to those integral to the interaction, whereby new species enter into the interaction. This type of convergence of species has occurred commonly in the evolution of mutualistic interactions, including those between pollinators (such as bees) and plants and those between vertebrates (such as birds and bats) and fruits.

Some of the species drawn into mutualistic interactions become co-mutualists, contributing as well as benefiting from the relationship, whereas others become cheaters that only exploit the relationship. In many interactions between bee pollinators and plants, bees collect the nectar from the reproductive parts of the plant and are often dusted with pollen in the process. When the bees fly to another plant of the same species, they may fertilize the plant by depositing pollen on the plant's stigma. In contrast, some bumblebees, such as those of Bombus terrestris, obtain nectar from the plant without picking up or dropping off pollen. They cheat by cutting through other parts of the plant instead of entering the flower.

Mutualism: yucca moth: In the mutualism between the yucca moth (Tegeticula yuccasella) and the yucca plant (Yucca), moth larvae feed on some—but not all—of the plants seeds and use the plant's seedpods as shelter. In return, adult moths serve as the plant's pollinator.

In other cases, the behaviour or appearance of several species may converge to enhance their mutual protection. For example, several species of heliconid butterflies that are distasteful to predators have evolved to resemble one another. In addition, one species may evolve to mimic the behaviour or appearance of another to garner some of the same protections enjoyed by the model species. This evolutionary strategy has been successful for nonvenomous snakes, such as the scar let king snake (Lampropeltis triangulum elapsoides), whose coloration closely resembles that of coral snakes, which can deliver a poisonous bite.

Coevolution is a complex process that occurs on many levels. It may appear in situations where one species interacts closely with several others, such as the interaction between European cuckoos (Cuculus canorus) and the other species whose nests they parasitize; it may involve many species, as in relationships between fruit-bearing plants and birds; or it may take place in some subgroups of species but not others. It is important to note that human activities often disrupt the process of coevolution by changing the nature and the extent of the interactions between coevolving species. Some examples of harmful human activities include habitat fragmentation, increased hunting pressure, and favouritism of one species over another and the introduction of exotic species into ecosystems that are ill-equipped to handle them.

Chapter 2

Evolutionary Biology and its Subfields

The branch of biology which deals with the evolution and evolutionary processes which are responsible for diversity of life on Earth is known as evolutionary biology. This chapter discusses the various subfields of evolutionary biology like evolutionary developmental biology, evolutionary ecology, evolutionary genetics, and population genetics. The topics elaborated in this chapter will help in gaining a perspective about the subfields of evolutionary biology.

Evolutionary Biology

Evolutionary biology is a branch of biology that is primarily concerned with the evolution of species. It encompasses other fields of biology such as genetics, ecology, systematics, and paleontology. Evolution pertains to the sequence of events depicting the gradual progression of changes in the genetic composition of a biological population over successive generations. These processes resulted in the diversity of life on earth and one of the fundamental tenets is that all life on earth evolved from a common ancestor (i.e. referred to as the last universal common ancestor). At present, research in evolutionary biology includes studies in molecular evolution, genetic drift, biogeography, etc. Some of the topics currently being studied are the evolution of sexual reproduction, the evolution of cooperation, speciation, evolvability, and so on. Recently, a sub-field emerged called evolutionary developmental biology. It is a field concerned with understanding the development of an organism from a single cell to its adulthood and finding out about the phylogeny among organisms. A person studying evolutionary biology is known as an evolutionary biologist.

Evolutionary Developmental Biology

Evolutionary developmental biology (evo-devo) is that part of biology concerned with how changes in embryonic development during single generations relate to the evolutionary changes that occur between generations. Charles Darwin argued for the importance of development (embryology) in understanding evolution. After the discovery in 1900 of Mendel's research on genetics, however, any relationship between development and evolution was either regarded as unimportant for understanding the processes of evolution or as a black box into which it was hard to see. Research over the past two decades has opened that black box, revealing how studies in evo–devo highlight the mechanisms that link genes (the genotype) with structures (the phenotype). This is vitally important because genes do not make structures. Developmental processes make structures using road maps provided by genes, but using many other signals as well—physical forces such as mechanical stimulation, temperature of the environment, and interaction with chemical products produced by other species—often species in entirely different kingdoms as in interactions between bacteria and squid or between leaves and larvae. Not only do genes not make structures (the phenotype), but new properties and mechanisms emerge during embryonic development: genes are regulated differentially in different cells and places; aggregations of similar cells provide the cellular resources from which tissues and organs arise; modules and populations of differently differentiated cells interact to set development along particular tracks; and organisms interact with their environment and create their niche in that environment. Such interactions are often termed "epigenetic," meaning that they direct gene activity using mechanisms that are not encoded in the DNA of the genes.

The Control of Body Structure

Deep Homology

Roughly spherical eggs of different animals give rise to extremely different bodies, from jellyfish to lobsters, butterflies to elephants. Many of these organisms share the same structural genes for body-building proteins like collagen and enzymes, but biologists had expected that each group of animals would have its own rules of development. The surprise of evo-devo is that the shaping of bodies is controlled by a rather small percentage of genes, and that these regulatory genes are ancient, shared by all animals. The giraffe does not have a gene for a long neck, any more than the elephant has a gene for a big body. Their bodies are patterned by a system of switching which causes development of different features to begin earlier or later, to occur in this or that part of the embryo, and to continue for more or less time.

The puzzle of how embryonic development was controlled began to be solved using the fruit fly *Drosophila melanogaster* as a model organism. The step-by-step control of its embryogenesis was visualized by attaching fluorescent dyes of different colours to specific types of protein made by genes expressed in the embryo. A dye such as green fluorescent protein, originally from a jellyfish, was typically attached to an antibody specific to a fruit fly protein, forming a precise indicator of where and when that protein appeared in the living embryo.

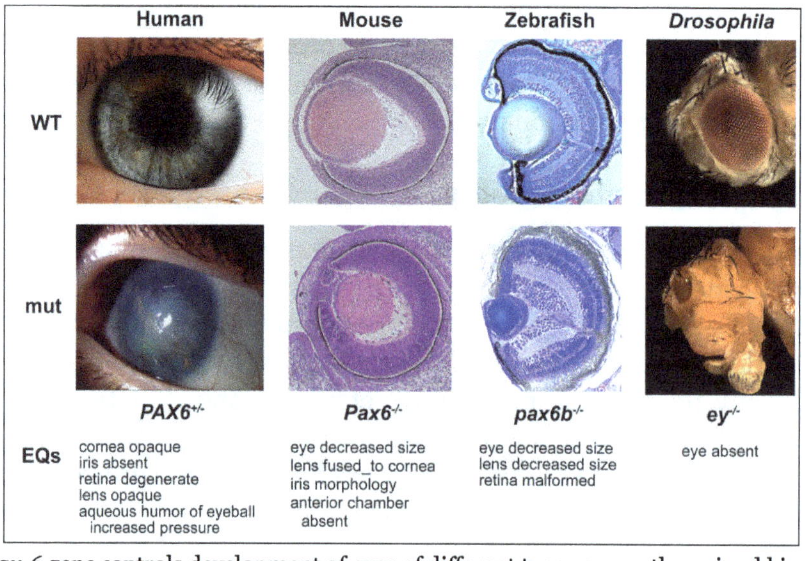

The *pax-6* gene controls development of eyes of different types across the animal kingdom.

Using such a technique, in 1994 Walter Gehring found that the *pax-6* gene, vital for forming the eyes of fruit flies, exactly matches an eye-forming gene in mice and humans. The same gene was quickly found in many other groups of animals, such as squid, a cephalopod mollusc. Biologists including Ernst Mayr had believed that eyes had arisen in the animal kingdom at least 40 times, as the anatomy of different types of eye varies widely. For example, the fruit fly's compound eye is made of hundreds of small lensed structures (ommatidia); the human eye has a blind spot where the optic nerve enters the eye, and the nerve fibres run over the surface of the retina, so light has to pass through a layer of nerve fibres before reaching the detector cells in the retina, so the structure is effectively "upside-down"; in contrast, the cephalopod eye has the retina, then a layer of nerve fibres, then the wall of the eye "the right way around". The evidence of *pax-6*, however, was that the same genes controlled the development of the eyes of all these animals, suggesting that they all evolved from a common ancestor. Ancient genes had been conserved through millions of years of evolution to create dissimilar structures for similar functions, demonstrating deep homology between structures once thought to be purely analogous. This notion was later extended to the evolution of embryogenesis and has caused a radical revision of the meaning of homology in evolutionary biology.

Gene Toolkit

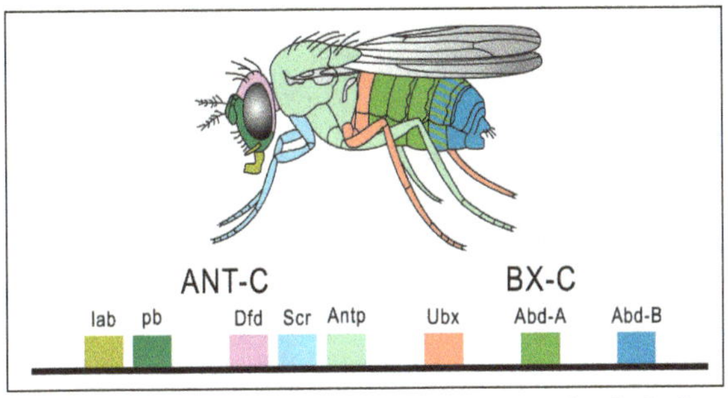

Expression of homeobox (Hox) genes in the fruit fly.

A small fraction of the genes in an organism's genome control the organism's development. These genes are called the developmental-genetic toolkit. They are highly conserved among phyla, meaning that they are ancient and very similar in widely separated groups of animals. Differences in deployment of toolkit genes affect the body plan and the number, identity, and pattern of body parts. Most toolkit genes are parts of signalling pathways: they encode transcription factors, cell adhesion proteins, cell surface receptor proteins and signalling ligands that bind to them, and secreted morphogens that diffuse through the embryo. All of these help to define the fate of undifferentiated cells in the embryo. Together, they generate the patterns in time and space which shape the embryo, and ultimately form the body plan of the organism. Among the most important toolkit genes are the *Hox* genes. These transcription factors contain the homeobox protein-binding DNA motif, also found in other toolkit genes, and create the basic pattern of the body along its front-to-back axis. Hox genes determine where repeating parts, such as the many vertebrae of snakes, will grow in a developing embryo or larva. *Pax-6*, already mentioned, is a classic toolkit gene. Although other toolkit genes are involved in establishing the plant bodyplan, homeobox genes are also found in plants, implying they are common to all eukaryotes.

The Embryo's Regulatory Networks

The protein products of the regulatory toolkit are reused not by duplication and modification, but by a complex mosaic of pleiotropy, being applied unchanged in many independent developmental processes, giving pattern to many dissimilar body structures. The loci of these pleiotropic toolkit genes have large, complicated and modular cis-regulatory elements. For example, while a non-pleiotropic rhodopsin gene in the fruit fly has a cis-regulatory element just a few hundred base pairs long, the pleiotropic eyeless cis-regulatory region contains 6 cis-regulatory elements in over 7000 base pairs. The regulatory networksinvolved are often very large. Each regulatory protein controls "scores to hundreds" of cis-regulatory elements. For instance, 67 fruit fly transcription factors controlled on average 124 target genes each. All this complexity enables genes involved in the development of the embryo to be switched on and off at exactly the right times and in exactly the right places. Some of these genes are structural, directly forming enzymes, tissues and organs of the embryo. But many others are themselves regulatory genes, so what is switched on is often a precisely-timed cascade of switching, involving turning on one developmental process after another in the developing embryo.

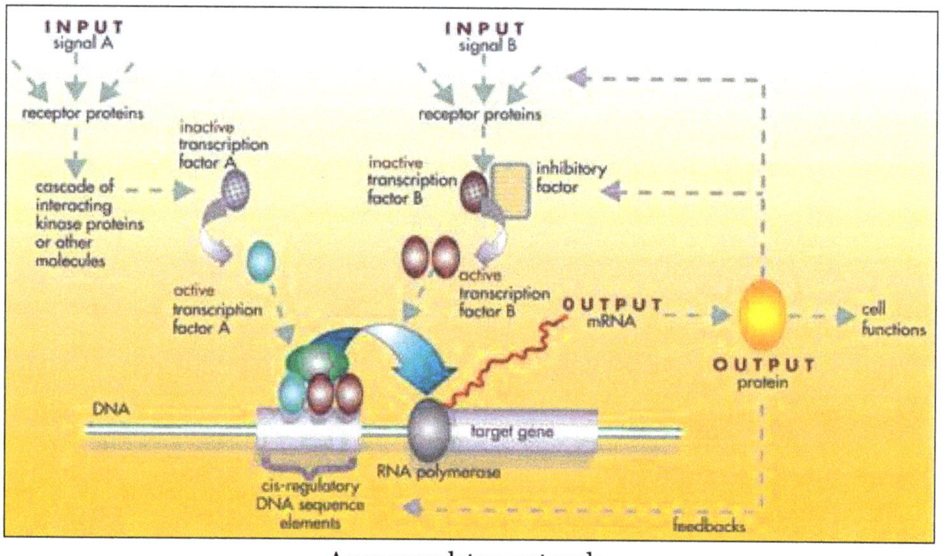

A gene regulatory network.

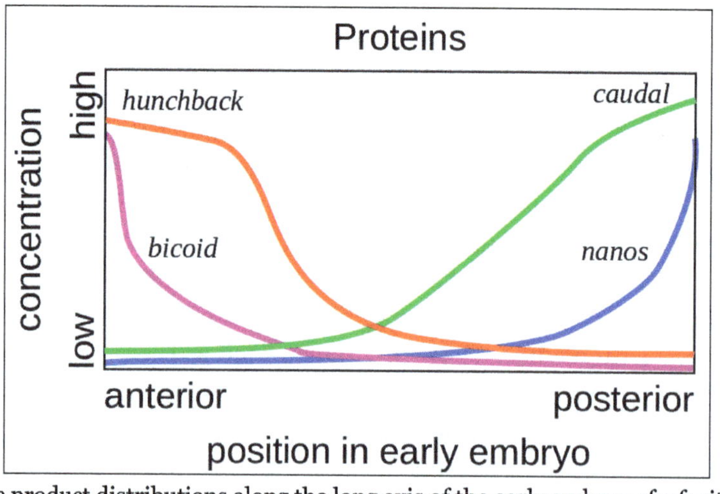

Gene product distributions along the long axis of the early embryo of a fruit fly.

Such a cascading regulatory network has been studied in detail in the development of the fruit fly embryo. The young embryo is oval in shape, like a rugby ball. A small number of genes produce messenger RNAs that set up concentration gradients along the long axis of the embryo. In the early embryo, the *bicoid* and *hunchback*genes are at high concentration near the anterior end, and give pattern to the future head and thorax; the *caudal* and *nanos*genes are at high concentration near the posterior end, and give pattern to the hindmost abdominal segments. The effects of these genes interact; for instance, the Bicoid protein blocks the translation of *caudal's* messenger RNA, so the caudal protein concentration becomes low at the anterior end. Caudal later switches on genes which create the fly's hindmost segments, but only at the posterior end where it is most concentrated.

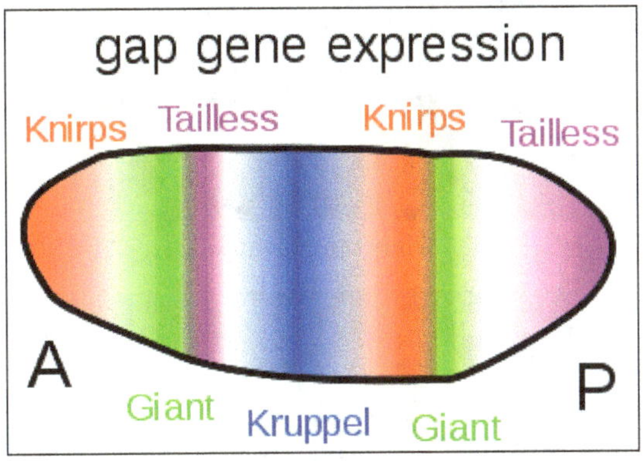

Gap genes in the fruit fly are switched on by genes such as bicoid, setting up stripes across the embryo which start to pattern the body's segments.

The Bicoid, Hunchback and Caudal proteins in turn regulate the transcription of gap genes such as *giant*, *knirps*, *Krüppel*, and *tailless* in a striped pattern, creating the first level of structures that will become segments. The proteins from these in turn control the pair-rule genes, which in the next stage set up 7 bands across the embryo's long axis. Finally, the segment polarity genes such as *engrailed* split each of the 7 bands into two, creating 14 future segments.

This process explains the accurate conservation of toolkit gene sequences, which has resulted in deep homology and functional equivalence of toolkit proteins in dissimilar animals (seen, for example, when a mouse protein controls fruit fly development). The interactions of transcription factors and cis-regulatory elements, or of signalling proteins and receptors, become locked in through multiple usages, making almost any mutation deleterious and hence eliminated by natural selection.

The Origins of Novelty

Among the more surprising and, perhaps, counterintuitive (from a neo-Darwinian viewpoint) results of recent research in evolutionary developmental biology is that the diversity of body plans and morphology in organisms across many phyla are not necessarily reflected in diversity at the level of the sequences of genes, including those of the developmental genetic toolkit and other genes involved in development. Indeed, as John Gerhart and Marc Kirschner have noted, there is an apparent paradox: "where we most expect to find variation, we find conservation, a lack of change". So, if the observed morphological novelty between different clades does not come from changes in gene sequences (such as by mutation), where does it come from? Novelty may arise by mutation-driven changes in gene regulation.

Variations in the Toolkit

Figure shows different species of *Heliconius* butterfly have independently evolved similar patterns, apparently both facilitated and constrained by the available developmental-genetic toolkit genes controlling wing pattern formation.

Heliconius erato

Heliconius melpomene

Variations in the toolkit may have produced a large part of the morphological evolution of animals. The toolkit can drive evolution in two ways. A toolkit gene can be expressed in a different pattern, as when the beak of Darwin's large ground-finch was enlarged by the *BMP* gene, or when snakes lost their legs as *distal-less* became under-expressed or not expressed at all in the places where other reptiles continued to form their limbs. Or, a toolkit gene can acquire a new function, as seen in the many functions of that same gene, *distal-less*, which controls such diverse structures as the mandible in vertebrates, legs and antennae in the fruit fly, and eyespot pattern in butterfly wings. Given that small changes in toolbox genes can cause significant changes in body structures, they have often enabled the same function convergently or in parallel. *distal-less* generates wing patterns in the butterflies *Heliconius erato* and *Heliconius melpomene*, which are Müllerian mimics. In so-called facilitated variation, their wing patterns arose in different evolutionary events, but are controlled by the same genes. Developmental changes can contribute directly to speciation.

Consolidation of Epigenetic Changes

Evolutionary innovation may sometimes begin in Lamarckian style with epigenetic alterations of gene regulation or phenotype generation, subsequently consolidated by changes at the gene level. Epigenetic changes include modification of DNA by reversible methylation, as well as non-programmed remoulding of the organism by physical and other environmental effects due to the inherent plasticity of developmental mechanisms. The biologists Stuart A. Newman and Gerd B. Müller have suggested that organisms early in the history of multicellular life were more susceptible to this second category of epigenetic determination than are modern organisms, providing a basis for early macro-evolutionary changes.

Developmental Bias

Among the centipedes, all members of the Geophilomorpha are constrained by a developmental bias to have an odd number of segments, whether as few as 27 or as many as 191.

Development in specific lineages can be biased either positively, towards a given trajectory or phenotype, or negatively, away from producing certain types of change; either may be absolute (the change is always or never produced) or relative. Evidence for any such direction in evolution is however hard to acquire and can also result from developmental constraints that limit diversification. For example, in the gastropods, the snail-type shell is always built as a tube that grows both in length and in diameter; selection has created a wide variety of shell shapes such as flat spirals, cowries and tall turret spirals within these constraints. Among the centipedes, the Lithobiomorpha always have 15 trunk segments as adults, probably the result of a developmental bias towards an odd number of trunk segments. Another centipede order, the Geophilomorpha, the number of segments varies in different species between 27 and 191, but the number is always odd, making this an absolute constraint; almost all the odd numbers in that range are occupied by one or another species.

Evolutionary Ecology

Evolutionary ecology examines the environmental factors that drive species adaptation. Studies of the evolution of species might seek to answer the question of how populations have changed genetically over several generations but might not necessarily attempt to learn what the underlying mechanisms might be. Evolutionary ecology seeks those mechanisms. Thus, in the well-known example of the peppered moth, the populations in the industrialized English Midlands changed over generations from having wings coloured largely grayish white, peppered with black spots, to wings that were mostly blackish. The ecological mechanism involved predation—birds readily detected the light-coloured moths against the background of the tree trunks that industrial pollution had darkened, whereas the dark-coloured moths remained generally undetected.

Evolutionary ecology also examines broader issues, such as the observations that plants in arid environments often have no leaves or else very small ones or that some species of birds have helpers at the nest—individuals that raise young other than their own. A critical question for the subject is whether a set of adaptations arose once and has simply been retained by all species descended from a common ancestor having those adaptations or whether the adaptations evolved repeatedly because of the same environmental factors. In the case of plants that live in arid environments, cacti from the New World and euphorbia from the Old World can look strikingly similar even though they are in unrelated plant families.

Light gray peppered moth (Biston betularia): A light gray peppered moth (Biston betularia) and a darkly pigmented variant rest near each other on the trunk of a soot-covered oak tree. Against this background, the light gray moth is more easily noticed than the darker variant.

Dark coloured peppered moth (Biston betularia): Against the background of a lichen-covered oak tree, a darkly pigmented peppered moth (Biston betularia) stands out, while the light gray moth (left) remains inconspicuous.

Evolutionary Genetics

Genetics plays a central role in evolutionary biology, because only heritable traits can evolve. Evolutionary genetics is the study of how genetic processes such as mutation and recombination are affected by the evolutionary forces of selection, genetic drift and migration to produce patterns of genetic variation at segregating loci and for quantitative traits, within and between populations and over short and long timescales.

When Charles Darwin developed the theory of evolution, he knew nothing about genetics. Hence, one of its biggest weaknesses was that Darwin had to base it on crude ideas of inheritance. Around the same time, Gregor Mendel discovered the laws of inheritance, but the scientific community initially failed to appreciate his work's importance. It was only in the 1930's that Dobzhansky, Fisher, Haldane, Wright, Mayr and others unified genetics and the theory of evolution in the 'modern synthesis'. Still, the modern synthesis was built on a basic understanding of genetics, with genes merely being particulate inherited information. The basics of molecular genetics, like the structure of DNA, were not discovered until the 1950's. When modern evolutionary psychology emerged from ethology and sociobiology in the late 1980's, it had a strong emphasis on human universals, borne from both the assumption that complex adaptations are monomorphic (or sexually dimorphic) and have to go back to at least the last common ancestor of all humans, and the methodological proximity to experimental cognitive psychology, which tends to treat individual differences as statistical noise. As a consequence, genetic differences between people were marginalized in evolutionary psychology. Evolutionary psychology and behavior genetics developed nearly orthogonally for over a decade. Behavior geneticists discovered that virtually every psychological or behavioral difference shows genetic variation and that the molecular genetic underpinnings of most heritable traits are far more complex than assumed in the modern synthesis. Meanwhile, evolutionary psychologists increasingly realized the importance of genetic variation, e.g. in models of sexual selection for attractiveness, intelligence and other assumed honest signals of genetic quality or heritable variation in life history traits. During the last decade, evolutionary genetics gradually gained acceptance among evolutionary psychologists, though most still defer fully incorporating the genetic perspective.

Evolutionary genetics is concerned with the mechanisms that explain the existence and maintenance of genetic variation in traits. All else equal, one would expect selection to deplete genetic variation in heritable traits related to fitness eventually. However, such genetic variation is ubiquitous and underlies stable individual differences that play prominent roles in psychological theories, be it as traits under intersexual (e.g. attractiveness, agreeableness, intelligence; and intrasexual selection (masculinity, aggressiveness; Puts, this volume), life history traits, formidability in recalibration theory, sociometer sensitivity, perceived vulnerability to infection in the behavioral immune system, attachment security, or the tendency to show strong reciprocity in cooperation. Though these theories ascribe adaptive roles to individual differences, more or less explicitly linking them to fitness, their genetic variation is often taken for granted.

Evolutionary genetics can help evolutionary psychologists unearth clues to the ultimate reasons behind e.g. humans' cognitive faculties that go beyond what can gleaned through paleontology and archaeology. This information can have very practical implications, e.g. helping to understand how natural and sexual selection, when altered through changing mores or policy, will affect certain traits.

Genetic Architecture

Some research in molecular genetics has been carried out with the aim of characterizing the genetic architecture of traits, sometimes also called the genotype-phenotype map. The genetic architecture of a trait can provide important clues to the evolutionary history and the mechanisms that govern the maintenance of genetic variation in the trait. Characterizing the genetic architecture of a quantitative trait would ideally involve its robustness to mutations (canalization) as well as its evolvability. It would also imply gauging its degree of pleiotropy (whether the genes involved also have simultaneous other effects) and the importance of non-additive genetic variation (i.e. epistasis and dominance, variation that does not breed true to the next generation). Unfortunately, many examinations of the genetic architecture are limited to estimates of the number and effect size of involved genetic variants. Often the goal in such examinations is predicting which molecular genetic studies will succeed in the gene hunt and lead to biological pathways and drug targets, not to discover the ultimate, evolutionary explanations for heritable variation in a trait.

It may feel like a step back from identifying causative genetic variants, but we feel it is prudent to set aside the exciting prospects of what a successful gene hunt might entail and the different ideas about how we might succeed at that, focusing instead on finding common theoretical ground.

Researchers disagree how, if ever, we might explain a substantial portion of the "missing heritability", the observable genetic variation left unexplained by molecularly identified genetic variants. The limits of currently available tools can sometimes act as blinders; so that some theoretically plausible genetic architecture is hidden in our blind spots. Fortunately, as rapid technological and statistical development in molecular genetics adds to our toolkit, fewer blind spots should impede us. Humility is still very appropriate though, considering fairly principal problems such as the sheer parameter explosion that is encountered when relating genomic sequences to traits.

Neither should we be too eager to jump to the conclusion that our purported core traits will be reflected at the genetic level. For example, Mitchell argued against the continuous liability-threshold model of psychiatric disease, saying that there truly are discrete disorders; we just tend to group them broadly and arbitrarily. Similar arguments can be construed for the structure of psychological traits like personality and intelligence.

In addition, there are often unresolved questions about the genetic architecture implied by the available evidence. For example, researchers used to believe that selection would reduce genetic variation in fitness traits, driving associated variants to fixation. This seemed to be borne out by low heritability coefficients. However, when researchers realized that fitness traits present a large target for mutation, they re-examined the same heritability data expressed as the mean-standardized coefficient of variation (an absolute measure) and obtained large estimates of genetic variation. Heritability expressed as a proportion of total variation (a relative measure) had only appeared small in comparison, dwarfed by the large environmental variation. The conceptualization of fitness traits effectively reversed through a more appropriate statistic for variation.

Our understanding of how the forces of evolution shape traits' genetic architectures will continue to evolve. Thus, we begin with mechanisms potentially maintaining genetic variation before we discuss methods to identify causative genetic variants.

Forces of Evolution

We will begin by introducing four basic forces that affect genetic variation in populations.

Mutation

All existing genetic variants once arose by mutation. Relative to the 6.4 billion base pairs of the human genomic sequence, mutations are rare events. Beneficial mutations are the rarest of all, the majority likely being neutral to fitness, with deleterious mutations making up the rest. Because the idea of a neutral mutation can be reduced to chance (or drift) being more important for its fate than selection, calling a mutation neutral also depends on its commonness, not just its effect size. A mutation with a small beneficial effect will have its fate determined mostly by chance while it is rare, because chance events can eliminate all copies. Once its frequency rises and in larger populations drift becomes relatively less important, so the mutation will be governed more by selection.

The most common mutational event in humans is the change of a single base pair (the letters of the DNA), but there are also deletions, duplications and insertions of base pairs or even longer parts of DNA (copy number variants). Aneuploidies (chromosomal aberrations), such as the duplication of chromosome 21 which causes Down syndrome, are rare but massive, accounting for most altered base pairs per birth. Except for aneuploidies, which are well-known to exponentially increase in frequency with advancing maternal age, all types of mutations occur more often on the paternal side, and increasingly so with advancing paternal age at conception. Proximately, this is often attributed to the continuous division of cells in the paternal but not maternal germline, but ultimate explanations such as Bateman's principle (male investment in each offspring is lower) should be kept in mind.

Selection

Selection occurs when there is heritable variation in fitness. Natural selection is frequently broken down into different subcategories. One grouping distinguishes positive, directional selection (favoring increases), disruptive selection, (favoring extremes), and stabilizing selection, (favoring decreased variation in a trait). Another grouping considers survival and sexual selection separately. Sometimes this is differentiated further into "episodes of selection". Survival selection could e.g. be divided into the chances of an ovum to be released in ovulation, sperm fertilizing an ovum, a zygote implanting, the pregnancy being carried to term, and surviving birth, to reproductive age, and further. Sexual selection might be divided into the odds of finding and attracting a mate, outcompeting same-sex rivals, the number of mates, the number of offspring per mate, and the fitness and number of offspring in the next few generations. Often the mistaken impression that selection has diminished in humans is, on closer inspection, limited to factors affecting perinatal and postnatal survival selection, with little heed paid to components of sexual selection.

Correlated Selection, Genetic Hitchhiking and Pleiotropy

Genetic variants are not independently selected for. As the term "genetic hitchhiking" vividly implies, alleles can hitch a ride on the coattails, or haplotype, of a neighboring allele that is being selected for or against. The chances of inheriting a specific gene from a parent are not independent from those of its neighbors because we inherit genes in chunks. Over generations, recombination breaks haplotypes apart. Long unbroken haplotypes signal strong recent selection for a new mutation, because the neighboring alleles of a beneficial mutation are "swept" along on the coattails before recombination can break them apart (known as a "hard sweep"). Shorter unbroken haplotypes can signal selection on standing (pre-existing) genetic variation. Two or more alleles that usually co-occur (are in "linkage disequilibrium") and thus form a haplotype can have different, even opposing effects on fitness. Until recombination breaks them apart, they cannot be selected for independently.

Alleles experience correlated selection not only through proximity. Even a variant at a single locus can have multiple, pleiotropic effects on fitness via different phenotypic consequences. It can also make sense to distinguish fitness effects of an allele in different episodes of selection. For example, a mutation may be selected for pre-meiotically in the testes, but lead to Apert syndrome later on.

Genetic Drift

Luck plays a lead role when numbers are small. If there are few carriers of even a highly beneficial genetic variant, random events can eliminate all of them. Similarly, a deleterious variant can be fixated by chance, or a beneficial rare variant can randomly get lost in recombination. Either way, a gene variant may drift to fixation or extinction just by chance. If all variants at a locus are common (because no single variant is infrequent and the population is large), the law of large numbers implies that it will take long before either drifts to fixation. In humans, a comparatively extremely low genetic diversity points to genetic bottlenecks having been an important instance of drift. Bottlenecks may occur through migration, such as when founder populations immigrated to North America, or when population sizes decreased dramatically through harsh conditions such as droughts, epidemics or ice ages. If the resulting population is small and not diverse (e.g. a clan), even beneficial alleles from the parent population may be lost through drift.

Gene Flow or Migration

When individuals carrying certain alleles move from one group to another, the frequency of alleles in each group also changes. This process is distinguished from unsystematic genetic drift, because relevant genetic variants may differentially influence the propensity to migrate and the success in each group and environment.

Maintenance Mechanisms

Prolonged directional or stabilizing selection on a trait will deplete its genetic variance. The mechanisms that maintain heritable variation in a trait can be understood as equilibria or trade-offs between the forces of evolution that change allele frequencies: selection, mutation, genetic drift, and gene flow. In some cases it may seem as if evolution should lead to alternative genetic architectures with fewer trades-offs. Note that evolution is not over and that optimal solutions may not always be sufficiently better to be selected over merely adequate ones, which is e.g. why we still have blind spots in our eyes.

First-Order Mechanisms

Mutation-Selection Balance (MSB)

Mutations continuously emerge. If they are entirely neutral, they are invisible to selection and may drift or hitchhike to extinction or fixation. But if they are deleterious, purifying selection will act against them. We rarely hear of dominant lethal mutations because they tend to be eliminated within one generation. Huntington's disease, which develops after the age of reproduction, is one example to the contrary.

If a trait is genetically complex, as most traits of interests to evolutionary psychologists likely are, many genes will be involved, not all of which play a crucial role. Hence, some deleterious mutations will be selected against less intensely and might linger for a few generations. If the mutational target size of a trait (the number of associated genetic loci) is large, mutations affecting the trait will accumulate, so that individuals carry a certain mutational load.

Thus, variation in a trait such as physical attractiveness can be maintained even though it is likely under directional selection. In research on the genetics of autism spectrum disorders, new mutations appear to explain about 15% of cases, though this should not be equated with the part that MSB plays for autism, which may well be larger owing to older, inherited mutation load. Debate revolves around the number of genes likely to be involved in a trait and on the question whether rare, recent or common, older mutations mostly disrupt such genes.

Mutations in Balance with Stabilizing vs. Directional Selection

Traits under mutation-selection balance can be meaningfully differentiated further. If increases in a trait are linked to increased fitness (directional selection), new mutations should usually cause a decline in the trait. This assumption is implicit in most studies of MSB.

If fitness is instead linked to a certain optimum in a trait, it is said to be under stabilizing selection. Stabilizing selection acts to increase robustness to deleterious mutations, for example by increasing genetic redundancy. For sexually recombining species, such as ours, it has also been suggested that increased mutational robustness need not imply a decrease in the evolvability of a trait (its potential to react to selection): redundancy reduces the selection pressure on individual variants and thus allows variation to build up in the back-up copy, creating a playground for genetic innovation. In this case, new mutations should cause comparatively smaller deviations from the optimum and might lead us to miss genetic associations if we focus on directional declines. The optimum would be expected to be the mean of a trait, at least in traits that were not subject to recent environmental changes. The shape of the eye might be an example of this exception: myopia (short-sightedness, elongated eyes) is more common than hyperopia (early-onset farsightedness, shortened eyes), but the preponderance of myopia sufferers might be attributed to changes in our environment, in which near work became common and time outdoors decreased. To determine the not immediately visible optima of psychological traits, researchers could draw on associations of trait levels with survival and mate preferences as proxies of fitness consequences.

Balancing Selection

We now introduce a class of balancing mechanisms. In all of them, one selection pressure is counteracted by another in a different location, time, developmental stage, social environment or intra-individual genetic context.

By Spatial Environmental Heterogeneity

Humans can experience different selection pressures in different environments. Selection by location need not be limited to selection pressures such as varying solar intensity or altitude, though these examples are best-characterized.

Because personality may affect one's penchant for travel, migration can support spatial balancing selection: If those who want to see the world keep leaving their home island for the mainland, the remaining islanders may end up less open to experience on average. Selection would also reduce variance in openness if sedentary islanders did not occasionally interbreed with visitors from the mainland. This sort of recurring gene flow can maintain variation in openness. Similarly, sociability supports migration tendencies from rural to urban areas. In scenarios such as these genetic variation is maintained because people within a population select themselves into the environments for which they are best adapted. Such niche picking (also known as active gene-environment correlation) is potentially a strong force in the maintenance of genetic variation in humans. In the population as a whole, no trait or underlying genetic variant would effectively be favored, thus the selection pressures would balance.

Because cultural and other environmental explanations are hard to disentangle from genetically based psychological differences between populations, we advocate a cautious approach to this controversial topic. Some jump to premature conclusions about major genetic differences and even superiority based on flimsy evidence such as fairly high within-group heritability coefficients, but a balanced view of the evidence shows how difficult explaining group differences genetically is.

Because of humans' ecological dominance and concomitant capacity to shape the environment to their needs (niche construction), Penke, Denissen, and Miller argued that the most important fluctuating aspect that humans need to adapt to is their social environment.

By Social Environment (Negative Frequency-Dependent Selection)

There are three morphs (types) of male common side-blotched lizards (Uta stansburiana), and three alleles at one Mendelian locus govern their throat color and concomitant behavior. Blue-throated males guard one mate and territory. Their mates can be stolen by larger, aggressive, orange-throated males, who keep large territories and multiple mates. Because they do not guard their mates well, they are vulnerable to having their mates stolen by yellow-throated males, who pretend to be female to sneakily gain access. This non-transitive mating game has been compared to rock-paper-scissors and leads to oscillations in which the least common morph becomes more common in the next generation.

Biological sex is probably the most familiar morph under such negative frequency-dependent selection (NFDS) in humans, as the rarer sex becomes more desirable and thus has reproductive advantages due to mating market forces. NFDS has also been invoked to explain primary psychopathy, personality traits and, perhaps most fruitfully, immunity to parasites.

If psychopathy were under frequency-dependent selection, we might, through altered policy, lower the equilibrium frequency of psychopaths within few generations.

Over Time (Generations)

If selection fluctuates over time more quickly than is needed for trait alleles to be driven to either fixation or extinction, variation can be maintained in oscillations. For example, if sex ratios in populations naturally fluctuate over time, genetic variation in personality traits that lead to better mating outcomes in one sex can be maintained by balancing selection. If the fluctuations are predictable, selection should act to create genetically fixed conditional (facultative) strategies instead, a rich topic for life history theory.

Over Time (Ontogenetic Development)

Earlier we mentioned an allele that proliferates in the testes, but leads to disease. Negatively correlated selection across developmental stages is also plausible for quantitative traits. For instance, large heads may support cognitive ability in later life, but they complicate birth. Selection should favor traits that are not subject to such trade-offs, but especially in conjunction with fluctuations of the fitness effects at different developmental stages, variation could be maintained.

By Genetic Variant at other Loci (Epistasis)

An allele may have a beneficial or deleterious effect only in the presence or absence of other genetic variants. The sheer complexity of considering all the interactions in conjunction with the already large number of variants in the human genome has led some to propose that evolution would lead to mainly additive and even modularized variation in certain traits, but epistasis might also be missed owing to insufficient statistical power.

By Genetic Variant at the Same Locus (Overdominance, Heterozygote Advantage, Selection-Drift Balance)

Consider a polymorphism, such as the one involved in sickle-cell anemia. Two copies of the polymorphism make blood cells sickle-shaped under lowoxygen conditions and typically lead to premature death. But having only one copy (heterozygosity) confers greater resistance to malaria. Individuals from areas in which malaria was a strong selective pressure are more often carriers of the sickle-cell polymorphism. Heterozygotes have a selective advantage over homozygotes with either allele and so the sickle-cell allele can persist in the population at equilibrium frequency.

These equilibria are not stable: an allele that has the benefits but not the disadvantages will easily displace its competitor. We expect to see overdominance especially under strong, recent selection, such as that incurred by epidemics.

Mutation-Drift Balance (Selective/Ancestral Neutrality)

If mutations affecting a neutral trait arise so frequently that some linger before they drift out of existence, we expect genetic variation in this trait to linger as well. Because of the nature of genetic drift, existing, entirely neutral polymorphisms would linger longer in large populations. Because most human DNA is nonfunctional junk, which is not conserved through purifying selection, most mutations are neutral. One's first intuition might then be that most human individual differences are selectively neutral or "evolutionary noise". However, a commonly variable trait that is phenotypically visible to selection is less likely to be entirely neutral. This is especially the case since we tend to be interested in traits because they have predictive value for consequential life outcomes such as reproductive success, and thus evolutionary fitness.

Additionally, because populations are larger nowadays, selection is more efficient, and will usually be stronger than drift.

In humans, with their rapidly changing culture and environment and with their rapidly increasing population size, we might want to pay special heed to traits that used to be selectively neutral or nearly so, but no longer are. These are traits where we might expect natural selection to rapidly deplete genetic variation. Because traits under mutation-drift balance have a repository of standing variation and because selection is stronger than drift, it can decrease previously maintained variation.

A potential candidate for an ancestrally neutral psychological trait may be our preference for rising early or late: Our circadian rhythm is entrained to a universal source of light, the sun, in areas with little artificial light. With more artificial light, individuals' circadian rhythms become more variable, and such differences are moderately heritable. Possibly what we see here is cryptic genetic variation, revealed only under artificial light. Without it the lack of variation in light exposure within populations might have meant that heritable differences were not visible, even though psychological differences that would have influenced self-exposure to artificial light already existed.

Mechanisms Implicating More than One Trait a Time

Mechanisms that lead to the impression that there is heritable variability in a trait, but which are best understood in conjunction with other mechanisms and traits.

Mechanisms Related to Pleiotropy and Hitchhiking

When genes are pleiotropic (affect multiple traits) or in linkage (in close proximity to each other on a chromosome), genetic correlations among traits can appear. There are ways to discover genetic correlations and to analyze contemporary selection on multiple correlated traits, but few studies have tried to do so for human evolutionary history.

The best-characterized examples of antagonistic pleiotropy arise in conjunction with biological sex. Traits like facial masculinity may be more adaptive in one sex than the other, but the respective alleles spend half their careers in each sex. Another important class of pleiotropic interactions may arise through the body's limited energy budget, especially that available for immune, brain and gut functions. As a consequence, selection cannot optimize either trait, eventually resulting in a continuum of equally fit trait combinations maintained in the population.

Reactive Heritability

Not every trait with heritable individual differences needs to be subject to some sort of balancing mechanism itself. Instead, it could be calibrated to another heritable trait. For example, Lukaszewski and posited that extraversion might be calibrated to one's physical attractiveness and strength. Hence, we would find the signature of mutation-selection balance when studying extraversion in isolation, but would come to different conclusions when examining developmental and situational calibration of extraverted behaviour to one's relative strength and attractiveness.

If they are not fixed at birth, we should not presume the primacy of physical traits. For example, we know that myopia appears to be linked to the amount of time children spend outdoors, but the substantial heritability estimates for myopia have led some researchers to downplay environmental explanations for the recent increase in myopia incidence. But if myopia heritability is partly reactive to children's heritable proclivity for outdoor play and if some children spend less time outdoors in recent times, which is plausible, these findings could be reconciled.

The Evolutionary Genetics Toolkit

These tools were assembled from both quantitative and molecular genetics, as well as evolutionary psychology. We will note what these tools can be used for, and how they are sometimes misused, but acknowledge how all of these methods make their contributions.

Twin and Family Studies

Twin studies are one of the oldest tools available and have withstood the test of time. They rely on the key difference between monozygotic (identical) twins and dizygotic (fraternal) twins: identical twins share all of their genes, while fraternal twins share on average half of the genes that were variable between their parents. A central result from twin studies is usually a heritability estimate, though the rich data from twin and family studies can answer many other questions too. The concept hails from plant and animal breeding, where it is used to predict response to artificial selection.

Estimates of heritability derived from twin studies have held up remarkably well when re-examined using different family relationships (e.g. parents, siblings, half- and adopted siblings) and can be easily extended to novel data such as the sometimes numerous offspring of sperm donors. In cases where selection is fairly clear-cut, estimates of heritability have borne out their usefulness as predictors of the response to selection. For example, children of sperm donors are taller in a manner consistent with their mothers' selection on donor height.

Usually things are not so tidy: Heritability estimates from twin studies often include some non-additive variation, i.e. variation that will not "breed true" to the next generation. Moreover, environmental confounds can make it hard to isolate an effect of selection, as the initiators of the Scottish Mental Survey discovered in 1947 when they attempted to show a decline of intelligence through differential fertility and found an increase instead Humans simply do not behave like crops on a field or cattle in a breeding facility; they actively choose mates and both choose and modify their environments. This decreases the value of heritability estimates as more than a proof that genetic differences play a role in observable phenotypic variation.

High heritability in twin studies has often been misunderstood to imply that a trait cannot be changed. To the contrary, species-typical universals such as two-leggedness have virtually zero heritability, because the underlying genes rarely vary. On the other hand, some gene-environment interactions were not apparent before the relevant environment changed: For example, developing phenylketonuria, a disease causing intellectual disability, depends on consuming phenylalanine, which was a universal part of our diet before its damaging effects in some individuals became known.

Linkage Studies

Linkage studies, which identify larger genetic segments that segregate according to disease status in a pedigree, have been useful tools in the identification of "simple" Mendelian disorders, where single genes have major effects. They might also help once we learn to tell apart phenotypically similar diseases that we now group as complex psychiatric disorders. Linkage studies for most psychological variation have been characterized as a let-down. Still, they ruled out a suggested genetic architecture: If there were, for example, a single genetic locus causing human psychopathy (i.e. an exploitative social strategy) in analogy with the aforementioned sneaky side blotched lizard, linkage patterns would have led to its identification.

Candidate Gene Studies

Candidate gene studies look for the association of a specific genetic locus with the trait of interest. By hypothesizing which locus may be involved a priori, they avoid correcting for multiple comparisons and can thus use smaller samples than the similar, but exploratory genome-wide association paradigm. They have come under intense criticism, because of non-replications and general doubts whether there is sufficient theory to predict candidate genes.

Some recent studies, however, successfully employ candidate gene approaches, implicating candidate gene sets and apparently building on stronger theory than before. For example, W. D. Hill and colleagues reported and replicated an association of intelligence with variation in genes involved in one of the postsynaptic density complexes that have been implicated in cognitive functioning. Through pre-registration of candidate genes, researchers could easily end disagreements and distrust whether their studies deserve the label of confirmatory research and concomitant relaxation of false discovery rates. Unfortunately, this is seldom done.

Genome-Wide Association Studies (GWAS)

GWAS assess the status of individuals on around a million genetic loci across the genomes that are commonly variable in the population. While GWAS directly assess only around 0.033% of the human genome this way, linkage disequilibrium makes the assessed variants fairly exhaustive markers of common genetic variation, which is then related to the variation in the trait of interest. GWAS require large samples and have been early adopters of harsh significance thresholds to account for the number of multiple comparisons.

GWAS have been successful in the identification of some of the genes that matter for pigmentation, some medical disorders, height, and recently schizophrenia. Yet, for most psychological traits, especially normal variation, they rarely identified replicable associations. This is often framed negatively, but GWAS effectively ruled out genetic architectures involving few common variants of medium-to-large effects for all psychological traits studied this way so far. Some researchers have advocated ever larger samples in order to potentially identify huge sets of genetic variants with individually miniscule effect sizes, while others argue that theory predicts only effects of questionable practical relevance and that family-based designs are better-suited.

Using Sequenced Exomes and Genomes in Association Studies

Sequencing refers to identifying every single base pair in someone's genome, not just a few commonly polymorphic loci, as in GWAS. When sequencing is limited to protein-coding genes (ca. 1-2% of the whole genome), this subset is called the exome. The exome constitutes a more manageable amount of data, and has been considered promising for clinical variation. However, much of it is conserved between species and a lot of recent selection has operated on promoters outside the exome, making exome variation a less likely candidate for contributing to the genetic architecture of psychological traits in the normal range.

With the amounts of data generated by genome sequencing, entirely exploratory research would not be useful due to the sample sizes required to filter chance findings. Integrating prior knowledge, such as annotations on regions with a signature of recent selection or expression in the brain, or alternatively relying on summary indices of rare genetic variants, a direct operationalization of mutation load may make such data manageable.

Genomic Prediction and Genome-Wide Complex Trait Analysis (GCTA)

A method formerly used primarily to predict breeding value in domestic animals has recently become popular in human genetics under the name GCTA. The general method estimates distant relatedness (less than fourth cousins) between individuals in the general population on the basis of common genetic variants, as provided by GWAS. Unlike GWAS, this method does not identify individual important loci. Instead the distant relatedness is used to infer a heritability score akin to that known from twin studies, but based solely on molecular data. After many GWAS failed to identify loci associated with psychological traits, GCTA provided a means of showing that the genotype data was actually informative: it can validate heritability estimates and be used to enable marker-assisted breeding (though this application is unlikely in humans), even if it does not identify causative genes and hence provides no foothold to find biological pathways. A frequently raised objection is that GCTA heritability estimates might be spurious, driven by the resemblance of distantly genetically related individuals for non-genetic reasons, such as similar environments because of shared ancestry and migration history. Researchers working with GCTA acknowledge such confounds, the discussion revolves mostly about whether the corrections are sufficient.

Some researchers also doubt whether finding high GCTA heritability implies that the infinitesimal model of many common variants of tiny effect applies, especially when debilitating disorders are under study.

Paternal Age Effects

By sequencing and comparing the genomes of both parents and an offspring, Kong and colleagues convincingly demonstrated that the number of newly occurred single nucleotide variants in offspring can almost entirely be accounted for by the father's age at conception. Thus, paternal age can be used as a proxy variable to infer the effect of new mutations. To isolate this effect, the fact that human reproductive timing is not governed by chance has to be statistically controlled. Initially reported negative associations between paternal age and intelligence in the normal range have not been replicated in later studies. Controlling parental intelligence, an important predictor of reproductive timing, may account for some of the observed heterogeneity of effects. Employing sibling comparison designs also led to the disappearance of paternal age effects on intelligence, while a strong association with attention deficit hyperactivity disorder became visible only with sibling controls.

Properly isolated, paternal age effects can provide evidence for a trait being under mutation-selection balance. In addition, they can be useful to predict the effect of increasingly delayed reproduction in the industrialized world, net of potentially more modifiable aspects such as parental investment.

Genome and Exome Triplets and Quads

When the entire exomes or even genomes of parent-offspring trios are sequenced, it becomes possible to count new mutations, i.e. alleles that neither parent carried. By assessing which haplotype a mutation lies on, it is also possible to identify the parent of origin. Then, mutation counts can predict e.g. intellectual disability and recurring mutations can be used to zero in on causative genes.

Exome quads (both parents and two offspring) have been used in autism genetics. Using genome annotations, Iossifov and colleagues estimated which mutations interrupted genes. By also sequencing unaffected siblings whose genomes were recombined from a common parental pool, they could isolate the effect of having more disrupted genes. Studies on autism genetics tried to isolate the effect of new mutations from assortative mating by only considering families without a familial history of autism and through sibling comparisons. These molecular genetic studies corroborate earlier results of autism increasing with paternal age.

Inbreeding Depression and Outbreeding Elevation

Inbreeding depression refers to a fitness decrease in offspring of consanguinous unions. Consanguinous parents (second cousins and closer) and their offspring make up about 10% of the world's population, though their prevalence has been predicted to decline. Franssen reported a linear negative relationship between offspring mental ability and consanguinity ranging from second-cousin-marriages to incest. Such associations are confounded by many unobserved common causes. For example, lower parental education can, via lower mobility, increase the likelihood of marrying relatives and thus inflate estimates of inbreeding depression. The family history and cultural prevalence of consanguinity (e.g. in clans and castes) affect inbreeding coefficients too, so that estimates based on just two generations can be off.

Outbreeding elevation, also known as hybrid vigor or heterosis, refers to the increased phenotypic quality of the offspring of genetically more distant parents. This phenomenon is very familiar to plant and animal breeders. Mules may be the most iconic hybrids and hybrid maize the most frequently consumed. The vigor does not necessarily translate to evolutionary fitness: mules are valued beasts of burden, but frequently infertile. This is because too distant genetic relationships between parents can break up co-adapted gene complexes during recombination, hence breaking vital functions such as the ability to reproduce. A bit of both may have happened when modern humans and Neanderthals interbred. Hybrid vigor can also occur when inbreeding ends: Mixed-breed dogs have higher life expectancy than most purebreds proposed that urbanization and generally less sedentism led to decreased inbreeding and might be partial causes for the recent increases in height and intelligence in industrialized countries.

Runs of Homozygosity

Analogously to GCTA, which employs DNA-based subtle relatedness to validate twin studies' estimate of heritability, runs of homozygosity (ROH) are an attempt to characterize subtle inbreeding on a molecular level.

If long stretches of a diploid genome are homozygous, i.e. both strands of DNA have the same variants; we can infer that closely related individuals have bred. If many shorter stretches are homozygous, we can infer ancient relatedness. The genomic approach has the benefit that inbreeding over several generations can be characterized, though it is important to supplement this with knowledge of the history of endogamous marriage, founder effects and population bottlenecks.

Homozygosity appears to play a role not only in well-characterized recessive disorders such as cystic fibrosis, but also for traits like personality. Power and colleagues found a zero-to-slightly positive association between ROH burden and intelligence, which conflicts with (possibly more biased) pedigree-based estimates of inbreeding effects.

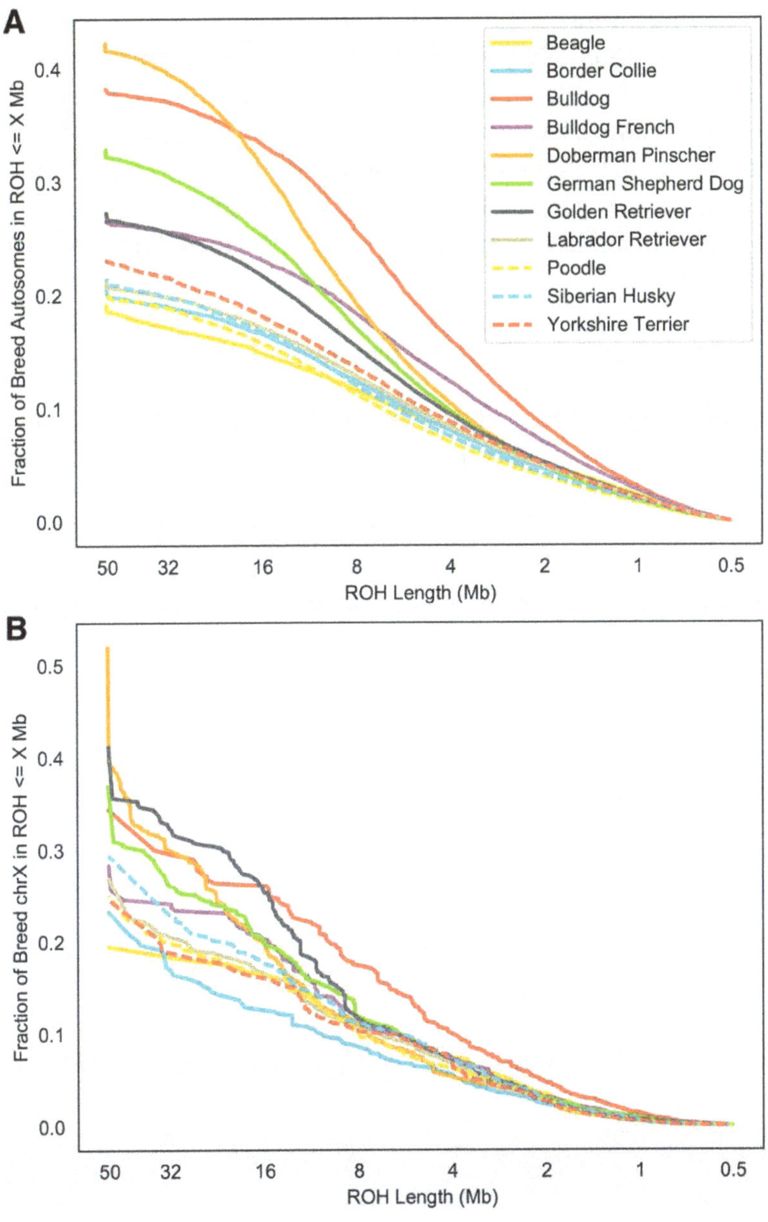

Cumulative density of runs of homozygosity (ROH) in 11 common breeds. A) Distribution of autosomal ROH across breeds. B) Distribution of chromosome X ROH in females only across breeds.

Relations with Fitness (Lifetime Reproductive Success) and Mate Preferences

This is because, with some exceptions (e.g. pervasive developmental disorders), it is difficult to establish that the same association has persisted over evolutionary time and is thus indicative of the balancing mechanism which primarily upheld variation in a trait. We lack historical data for psychological traits, and many associations between normal variation and fitness estimated nowadays could be fickle. Contemporary selection on human individual differences is interesting in itself, but we expect evolutionary genetics, among other disciplines, to answer the question "Why did humans evolve to be this way?"

In the age of widespread effective contraception it can be argued that mate preferences and choices are better-preserved indicators of sexual selection than correlations with reproductive success. In addition to being more immediately assessable than lifetime reproductive success, mate preferences have been shown to be relatively culturally invariant, unlike total fertility. Perinatal and postnatal survival selection plausibly have decreased in intensity since the advent of hygiene, modern health care, less frequent infanticide and lower infant and maternal mortality. Still, a large number of pregnancies are not carried to term and many debilitating, previously lethal genetic conditions, such as severe disability, may now be sexually selected against owing to lower attractiveness in the mating market.

Correlations with Indicators of Developmental Stability

Bilateral fluctuating asymmetry (FA) of the body is presumed to be an indicator of developmental stability, operating under the assumption that mutation-free organisms in good condition will be more symmetrical. Correlations with FA are thus assumed to provide an indirect way to tap a trait's association with mutation load. This paradigm is prevalent in evolutionary psychology and somewhat plagued by publication bias. Hardly any studies take a molecular or population genetic approach to fluctuating asymmetry in humans. Future studies should more directly examine an association of developmental stability indicators with rare genetic variant burden, paternal age, or consanguinity before correlations with FA can be deemed valid proxies for tapping 'good genes'. Pre-registration of studies could foster greater trust, especially that of scientists in adjacent domains such as genetics.

Chapter 3

Developmental Biology: Principles and Concepts

The subfield of biology which includes the processes of growth and development of plants and animals is known as developmental biology. This chapter encompasses developmental biology, human development, plant development and animal development. All the diverse principles of developmental biology have been carefully analyzed in this chapter.

It is the field of biology that studies the processes by which multicellular organisms grow and develop, controlled by their genes. It involves the study of mechanisms of development, differ entiation, and growth in animals and plants at the molecular, cellular, genetic and evolutionary levels. It also encompasses the biology of regeneration, asexual reproduction, metamorphosis, and the growth and differentiation of stem cells in the adult organism. The study of development has become essential for understanding any other area of biology.

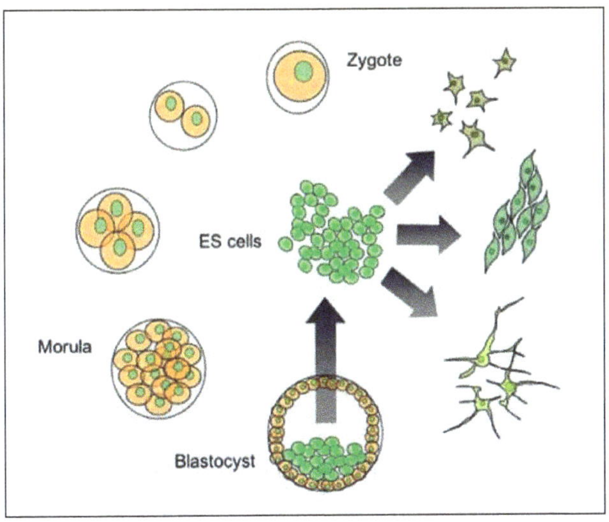

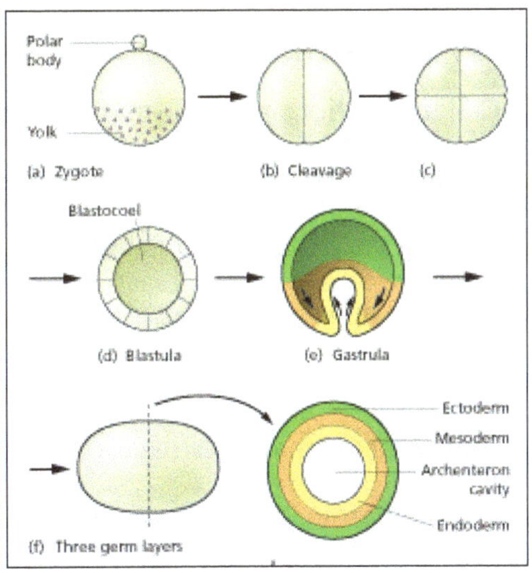

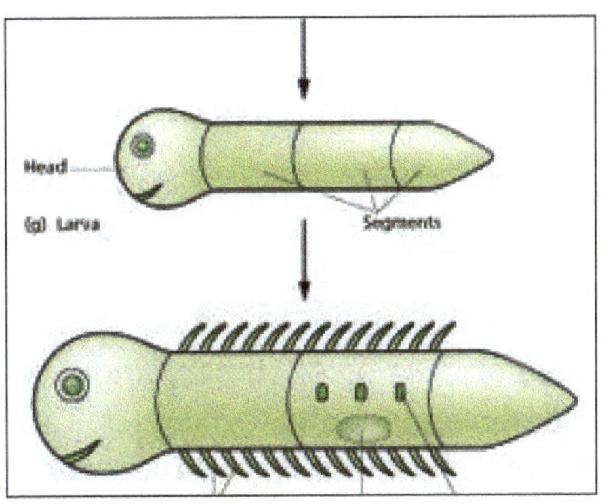

Interrelated Fields

It is one of the fastest-growing and most exciting fields in biology, creating a framework that integrates:

- Molecular biology
- Physiology
- Cell biology
- Anatomy
- Cancer research
- Neurobiology
- Immunology
- Ecology
- Evolutionary biology

Areas of Research

- Molecular genetics of development.
- Control of gene expression.
- Cell interactions and cell-matrix interactions.
- Mechanisms of differentiation.
- Growth factors and oncogenes.
- Regulation of stem cell populations.
- Evolution of developmental control.
- Gametogenesis and fertilization.

Other areas of particular emphasis include transcriptional control mechanisms, embryonic patterning, cell-cell interactions, growth factors, and signal transduction, and regulatory hierarchies in developing plants and animals.

Significance of Developmental Biology Studies

It helps to explain how a variety of interacting processes generate an organism's heterogeneous shapes, size, and structural features that arise on the trajectory from embryo to adult, or more generally throughout a life cycle.

It helps to understand the molecular, genetic, cellular, and integrative aspects of building an organism. Knowledge of normal developmental processes can aid in the understanding of developmental abnormalities and other conditions such as cancer. Philosophers of biology have shown renewed interest in developmental biology due to the potential relevance of development for understanding evolution and the theme of reductionism in genetic explanations.

Applications

Developmental Biology enquires about the fundamental processes that underpin the fertilization of an egg cell and its step-by-step transformation into the fascinating complexity of a whole organism which renders many applications in different fields.

In Fertility Clinics

Studying the underlying processes of fertilization has provided the foundations for much of what fertility clinics can do these days. Developmental biology investigates how fertilized egg cells divide in regulated manners to grow into full-size bodies, how the cells formed in this process communicate in meaningful ways to become different from each other, migrate, change shape and attach to each other, thus assembling into tissues and complex organs.

In Tissue Engineering

'Tissue engineering', which aims to grow replacement tissues in a plastic dish, is essentially guided by DB research. In cancer, cells lose their identity, divide excessively, detach from their local environments and migrate to form metastases. Much of this understanding that can instruct cures to contain these aberrant cells comes from DB research.

Tissues keep so-called stem cells which can be re-activated in orderly manners to divide and grow replacement tissues. There are high hopes from stem cell research, for example, to replace cartilage in arthritis or damaged discs, or brain cells in dementia, much of which is guided by the vast knowledge gained through developmental biology.

In Breeding Programs

For example, understanding plant development provides a means to speed up breeding processes, such as optimizing root systems, plant size or flowering time, thus contributing to the efforts of achieving sustainable food security in times of over-population.

In Understanding Factors Influencing Development

Understanding environmental influences on development, such as temperature-dependent sex determination in turtles, has enormous importance for conservation biology, especially in times of increasing pollution and global warming.

Plant Development

Plant development is a multiphasic process in which two distinct plant forms succeed each other in alternating generations. One form, the sporophyte, is created by the union of gametes (sex cells) and is thus diploid (contains two sets of similar chromosomes). At maturity, the sporophyte produces haploid (containing a single set of chromosomes) spores, which grow into the gametophyte generation. At their sexual maturity, the gametophytes produce haploid gametes that unite to begin a new cycle.

Figure shows the life cycle of the fern. (1) Clusters (sori) of sporangia (spore cases) grow on the under surface of mature fern leaves. (2) Released from its spore case, the haploid spore is carried to the ground, where it germinates into a tiny, usually heart-shaped, gametophyte (gamete-producing structure), anchored to the ground by rhizoids (rootlike projections). (3) Under moist conditions, mature sperm are released from the antheridia and swim to the egg-producing archegonia that have formed on the gametophyte's lower surface. (4) When fertilization occurs, a zygote forms and develops into an embryo within the archegonium. (5) The embryo eventually grows larger than the gametophyte and becomes a sporophyte.

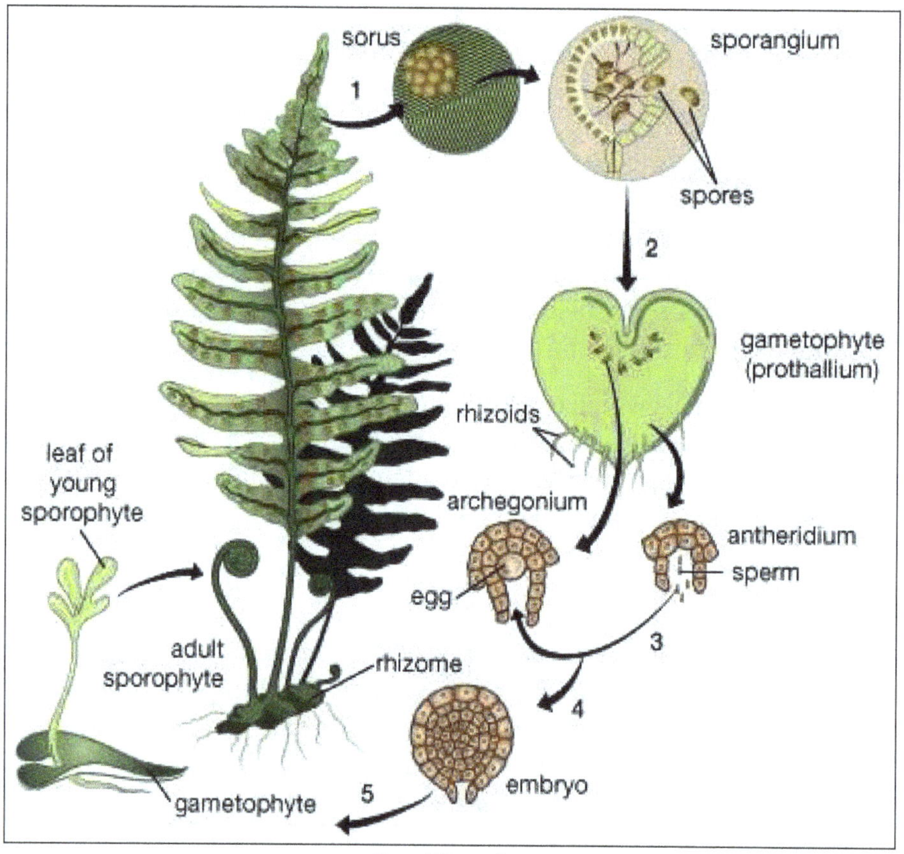

Fern life cycle.

Although both plants and animals share the chemical basis of inheritance and of translation of the genetic code into structural units called proteins, plant development differs from that of animals in several important ways. Higher plants sustain growth throughout life and, in this sense, are perpetually embryonic; animals, on the other hand, generally have a determinate period of growth, after which they are considered mature. Furthermore, both growth and organ formation in plants are influenced by their possession of a rigid cell wall and a fluid-filled space called the vacuole, two features unique to the plant cell. Conversely, certain features of animal cells are absent in plants. Notable is the lack of cellular movements and fusions that play an important part in tissue and organ development in higher animals.

Life Cycles

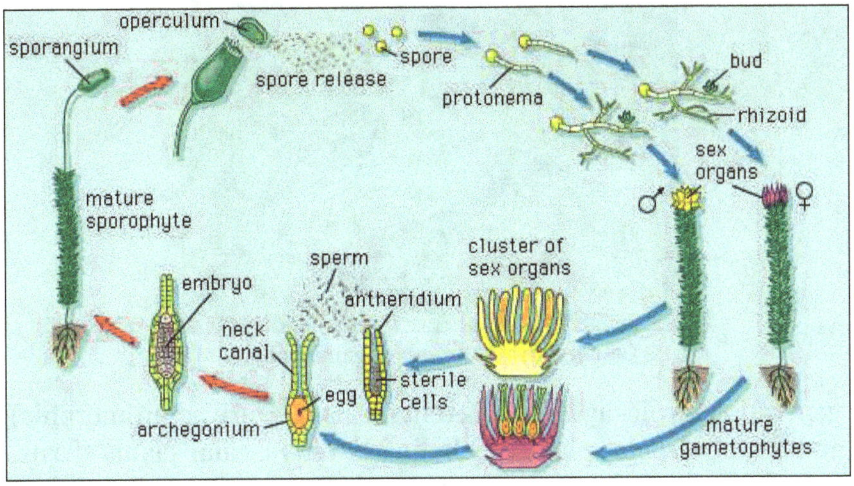

Life cycle of a moss.

The life cycle of all tracheophytes (vascular plants), bryophytes (mosses and liverworts), and many algae and fungi are based on an alternation of generations, or different life phases: the gametophyte, which produces gametes, or sex cells, alternating with the sporophyte, which produces spores. Gametophytes develop from the spores and, like them, are normally haploid; i.e., each cell has one set of chromosomes. Sporophytes develop from a fertilized egg, or zygote that results from the fusion of gametes (fertilization) formed by the gametophytes and is accordingly diploid; i.e., each cell has two sets of chromosomes. Although the two generations are phases of one life cycle, they have independent developmental histories; each begins as a single cell, passes through a juvenile period, matures, and gives rise to the alternate phase.

Figure shows the life cycle of bryophytes consists of an alternation of two stages, or generations, called the sporophyte and the gametophyte. Each generation has a different physical form. When a spore germinates, it usually produces the protonema, which precedes the appearance of the more elaborately organized gametophytic plant, the gametophyte, which produces the sex organs. The female sex organ is a flask-shaped structure called the archegonium. The archegonium contains a single egg in a swollen lower portion that is more than one cell thick. The neck of the archegonium is a single cell layer thick and sheathes a single thread of cells that forms the neck canal. When mature and completely moist, the neck canal cells of the archegonium disintegrate, releasing a column of fluid to the neck canal and the surrounding water. The egg remains in the base of the archegonium, ready for fertilization. The male sex organ, the antheridium, is a saclike structure made up of a jacket of sterile cells one cell thick; it encloses many cells, each of which, when mature, produces one sperm. When wet, the jacket of the mature antheridium ruptures to release the sperm into the water. When a sperm enters the field of the fluid diffused from the neck canal, it swims toward the site of greatest concentration of this fluid, therefore down the neck canal to the egg. Upon reaching the egg, the sperm burrows into its wall, and the egg nucleus unites with the sperm nucleus to produce the diploid zygote. The zygote remains in the archegonium and undergoes many mitotic cell divisions to produce an embryonic sporophyte. Mature bryophytes have a single sporangium (spore-producing structure) on each sporophyte. The sporangium generally terminates an elongate stalk, or seta, when the sporangium is ready to shed its spores and is capped by a lid, or operculum. The sporangium rupture usually involves specialized structures that enhance expulsion of the spores away from the parent gametophyte.

Gametophyte: A liverwort gametophyte.

The alternating generations typically have different forms (i.e., are heteromorphic); this is true for the bryophytes and for all vascular plants, including lower vascular plants (ferns and allies), angiosperms (flowering plants), and gymnosperms (conifers and allies). General rules for vascular plants are that the sporophyte generation is physically the larger, has a more complex developmental history, produces a greater range of cell types, and expresses a more diverse biochemistry; the gametophyte is often diminutive, reduced in the case of the angiosperms to a mere few cells. In the bryophytes, the gametophyte generation, rather than the sporophyte, is the more conspicuous.

Although the gametophyte generation in vascular plants is small and has limited physiological capabilities, its cells must convey genes capable of directing the sporophytic developmental pattern, because the pattern is transmitted through the gametes to the zygote. The expression of "sporophytic" genes must therefore be repressed in the gametophyte, probably from the time of spore formation (sporogenesis). Correspondingly, events associated with gamete formation (gametogenesis) or fertilization must somehow free the sporophytic genes and thus permit the zygote to enter the sporophytic developmental pattern. Although it might be supposed that the "switch" is associated with the difference in chromosome number between the haploid spore (a single set) and the diploid zygote (a double set), this has been shown not to be the determining factor.

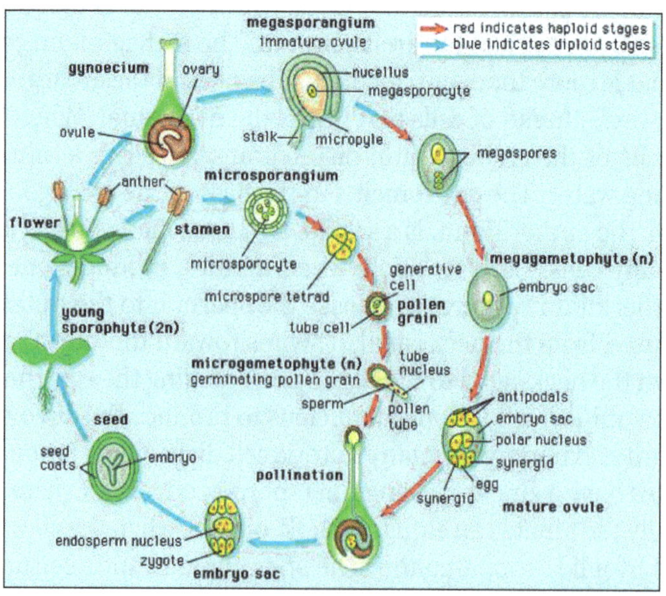

Angiosperm life cycle.

Figure shows life cycle of a typical angiosperm. The angiosperm life cycle consists of a sporophyte phase and a gametophyte phase. The cells of a sporophyte body have a full complement of chromosomes (i.e., the cells are diploid, or 2n); the sporophyte is the typical plant body that one sees when one looks at an angiosperm. The gametophyte arises when cells of the sporophyte, in preparation for reproduction, undergo meiotic division and produce reproductive cells that have only half the number of chromosomes (i.e., haploid, or n). A two-celled microgametophyte (called a pollen grain) germinates into a pollen tube and through division produces the haploid sperm. An eight-celled megagametophyte (called the embryo sac) produces the egg. Fertilization occurs with the fusion of a sperm with an egg to produce a zygote, which eventually develops into an embryo. After fertilization, the ovule develops into a seed, and the ovary develops into a fruit.

The alternation of generations illustrates an important principle, namely that cell lineages arising from single parental cells containing the same genetic potentiality may pursue mutually exclusive developmental patterns. Channelling, or canalizing, events of this nature occur repeatedly in the course of development of an individual plant, beginning with the pattern of cell division from the very first cleavage of the zygote cell.

Body Plans

Collectively, plants manifest a wide range of body plans, ranging from small multicellular structures to enormous trees. Among nonvascular plants, true parenchyma is found in the bryophytes, in both the gametophyte and sporophyte phases. The development of the moss gametophyte illustrates the transition from a filamentous to a highly organized three-dimensional growth form. The moss spore germinates into a filamentous plant, the protonema, which later produces a leafy shoot. This type of transition from simple to more complex growth form is accompanied by the synthesis of new kinds of ribonucleic acids (RNA's), presumably through the activation of genes that were not expressed during the early growth of the gametophyte.

Much of the remainder of this section is concerned with the development of the complex body forms of vascular-plant sporophytes, which do not normally pass through any filamentous stages. It may be noted, however, that, in the course of evolution, the capacity for this type of growth has not been lost, since it may be adopted by cells grown in tissue cultures in the laboratory.

Preparatory Events

The sporophytes of all vascular plants produce cells called spore mother cells—since they will give rise to spores—in spore cases (sporangia). Spore mother cells are usually surrounded, during development, by a special nutritive tissue. In the more primitive groups, each sporangium holds many mother cells. This is true also in the pollen-producing sporangia of gymnosperms and angiosperms but not in the egg-producing sporangia (ovules), which usually have only one mother cell.

In certain lower vascular plants, typified by the spike moss Selaginella, the gametophyte is formed entirely—or almost entirely—within the spore wall. Two kinds of gametophytes develop from the two kinds of spores produced by the sporophyte in different sporangia; the larger spore (megaspore) gives rise to the female gametophyte, the smaller spore (microspore) to the male. This condition is referred to as heterospory. The gametophytes, or prothalli, of other club mosses and most horsetails and ferns are sexually undifferentiated and arise from one kind of spore, a condition termed homospory.

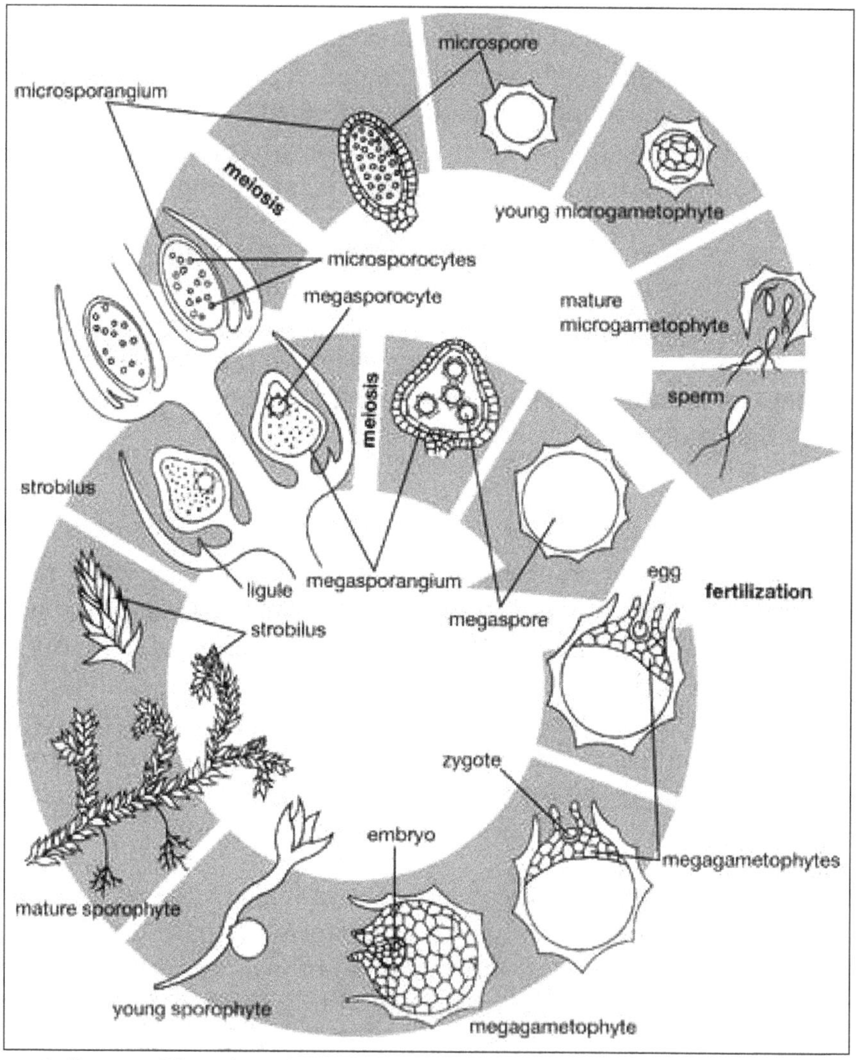

Spike moss life cycle: Life cycle of Selaginella, the only genus of spike moss.

In these homosporous groups, the gametophytes develop as free-living and independent plants that ultimately produce the gametes. In general, the male gametes (antherozoids) are produced in globose structures (antheridia) that are either stalked or sunken in the gametophyte. The antherozoids, always many in number, develop from mother cells enclosed in the jacket of the antheridium. Each antherozoid can move by using its whiplike hairs, or flagella, two or three (in the lycopods) or many (in the horsetails and ferns). The female gametes are formed singly in flask-shaped structures (archegonia) that also are either stalked or sunken in the gametophyte. The neck of the flask is closed by neck canal cells, which later break down to permit the entry of the male gamete. The egg itself lies in the basal part, or venter, of the flask, with a ventral canal cell above it. When the male gametes, or antherozoids, are released by the rupture of the antheridium, they swim in a water film to the archegonia and effect fertilization.

Wind pollination: Windblown pollen from the male cone of a lodgepole pine (Pinus contorta).

Among the gymnosperms the male gametophyte is much reduced and is a parasite on the sporophyte for only a short time. Cell cleavages within the spore wall cut off a prothallial cell, which will give rise to the vegetative (i.e., non-reproductive) part of the plant, and an antheridial cell, which divides into a tube cell and a generative cell. The male gametophyte so formed and contained within the spore wall is the pollen grain. After transfer to the ovule by wind, the pollen grain germinates to form a tube, and the generative cell divides into two cells, one of which forms the male gametes by further division. The gametes bear numerous spirally arranged flagella. The female gametophyte meanwhile develops entirely within the parent sporangium in the ovule. The size of the single functional spore increases greatly as the spore nucleus divides repeatedly to produce numerous free nuclei. Cell-wall formation then begins at the periphery, extending inward until the whole area is divided into cells. Up to four archegonia are formed, sunken in the tissue of the gametophyte, each with a female gamete, or egg.

The end of the gametophyte phase and the beginning of the sporophyte phase occur at fertilization, when one of the male gametes fuses with the female gamete to form the zygote, which will then develop as the sporophyte. (Development of the sporophyte can, in some cases, be triggered by means other than fertilization, in which case the organism is said to arise parthenogenetically.)

The male gametophyte of angiosperms is reduced to three cells, one so-called vegetative cell and two male gametes. The division producing the gametes may occur either before dispersal of the pollen grain or later, during the growth of the pollen tube. The female sporangium has one or two coats, or integuments, except for an opening (micropyle) at one end; the sporangium with an integument is called the ovule. The female gametophyte, known in this group as the embryo sac, develops from the parent spore while it is still retained in the sporangium. Three cell divisions result in eight nuclei, which arrange themselves so that three lie at each end and two lie in the centre. The cytoplasm then cleaves and three cells are formed at each pole, leaving two nuclei in a large central cell. The three cells at the micropylar pole (end toward the micropyle) form the egg apparatus. Two of these cells, called synergids, correspond to the neck cells of an archegonium; the third is the egg cell. The three cells at the opposite pole, the antipodals, play a part in embryo nutrition in certain genera. The two polar nuclei in the central cell ultimately unite, becoming the fusion nucleus. The pollen grain is transferred by various agencies (wind, water, animals) to the stigma of the female flower, and, as in the gymnosperms, it germinates to produce a tube. This tube grows through intervening tissues, through an opening (micropyle) of the egg, and enters a cell near the micropyle (synergid), in which the two male gametes are discharged. The unique feature of this phase of angiosperm development is that two fertilizations occur. One male gamete fuses with the egg to give the diploid zygote; the other makes its way to the fusion nucleus in the central cell, already diploid, and by a second fusion gives a triploid primary endosperm nucleus, which is later concerned in the formation of the nutritive tissue, or endosperm.

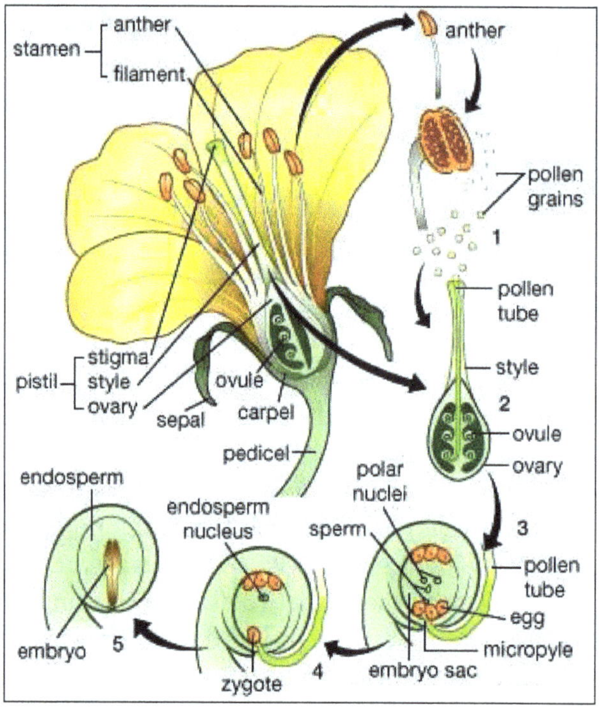

The life cycle of a flowering plant.

Figure shows (1) A pollen grain is released from the anther and settles on the stigma. (2) A pollen tube forms and grows through the style toward the ovule opening (micropyle). (3) Two of the nuclei (polar nuclei) in the ovule's embryo sac migrate to the centre to form a single cell. Three cells migrate to the micropyle, and one enlarges to become the egg. Two sperm that have formed from mitotic division of the pollen grain's generative cell enter the embryo sac through the micropyle. (4) One sperm fuses with the egg, resulting in a fertilized egg (zygote), which develops into an embryo. The second sperm fuses with the two polar nuclei to form the endosperm nucleus. (5) This nucleus divides to form a tissue (endosperm) that provides nutrients for the developing embryo.

Early Development: From Zygote to Seedling

Embryo Formation: Cleavage of the Zygote

In vascular plants embryo formation, or embryogenesis, usually occurs within a few hours after fertilization, with the first cell division that cleaves the zygote, or fertilized egg, into two daughter cells. Thereafter, rapid cell division provides the building blocks of the primary organs of the embryo sporophyte: the first root, first leaves, and the shoot apex. Temporary structures concerned with embryo nutrition—suspensor and foot—may also be produced. These organs originate in a polarization established at the time of zygote cleavage, but the details of their development vary widely among the different groups.

In the club mosses the zygote divides in a plane at right angles to the axis of the archegonium. The daughter cell toward the neck forms a short filament of cells, the suspensor; the inner cell gives rise to the other organs of the embryo, the shoot, root, and foot. The axis of the embryo is inclined to that of the archegonium and may be almost at right angles. This is in contrast to the behaviour of the true mosses, in which the embryo is oriented along the length of the archegonium, with the foot directed inward and the structures that are equivalent to the shoot, namely the spore capsule and its stalk, directed toward the neck.

A polarity like that of the mosses appears in the horsetails, in which the zygote divides by transverse and longitudinal walls to form a group of four cells. Of these, the two cells toward the neck give rise to the shoot system; the inner two produce the foot and root.

The details of early embryogenesis in gymnosperms vary considerably. In the cycads and ginkgos, the initial cleavage establishes a polarity opposite of that in the horsetails, the inner cell giving rise to the shoot and the outer producing the root. Many conifers are unique in that the zygote undergoes a period of free-nuclear division without cell formation, producing usually four or eight nuclei, which move to the end of the zygote, away from the neck cells, where cleavage begins. In the pines a further division gives four tiers of four cells. The intermediate tiers extend greatly to form a suspensor; each of the four cells at the lower pole may act as the parent cell of an embryo, a condition sometimes referred to as polyembryony.

In contrast, there is no free-nuclear stage in angiosperm embryogenesis. The zygote cleaves by a wall more or less at right angles to the axis of the embryo sac. The daughter cell next to the micropyle (basal cell) produces a suspensor and contributes to the root; the inner (terminal) cell gives rise to the shoot system. (Angiosperm embryogenesis is more fully described in the following section dealing with the origin of primary organs.)

Notwithstanding the variation in the different groups, the pattern of development established in the early cell cleavages is consistent. The primary polarization of the zygote must necessarily be imposed by the adjacent tissues of the sporophyte, but thereafter the fate of daughter cells depends on control established within the young sporophyte itself.

Although it is often possible to specify the origin of the cell lineages contributing to the various organs and tissue layers, a geometric regularity in cell division is generally maintained through only the first few division cycles in the embryo. The final form of the embryo is thus determined not through the specification of a precise scheme of cell division, as in the development of colonial algae, but through an overall control in which cell and tissue interactions play an important part.

Origin of the Primary Organs

Angiosperm embryogenesis can be described in terms of a much studied flowering plant called shepherd's purse (Capsella bursa-pastoris). The zygote divides into two cells, the terminal cell and the basal cell. The terminal cell divides by a wall formed at right angles to the first cleavage wall and then again by a wall formed at right angles to this; a quadrant of cells is thus formed. The partition of the quadrant cells in a transverse plane then produces an octant stage. By transverse divisions, the basal cell forms a filament, the suspensor, of up to ten cells, the end cell of which swells to form an absorbing organ. The attachment cell, or hypophysis, adjoins the octants derived from the terminal cell.

At this time, the prospective future of each of the zones of the embryo can be specified. Four cells of the octant group will ultimately produce the seed leaves (cotyledons) and the shoot apex; the other four will form the hypocotyl, the part of the embryo between the cotyledons and the primary root (radicle). The hypophysis will give rise to the radicle and the root cap; the cells of the suspensor will degenerate as the embryo matures.

The zones of the embryo destined to form the principal organs are established by this first sequence of divisions, and tissue layers are defined during the ensuing divisions. The octant cells divide by curved walls parallel to the surface; in this way the outer layer responsible for producing the epidermis of the shoot system is defined. Divisions of a more irregular nature in the inner zone ultimately define the tissues from which the central cylinder and vascular core of the main axis of the shoot will develop. Simultaneously, the hypophysis forms a group of eight cells by three successive divisions, the planes of which are mutually at right angles. Of these eight cells, the outer four produce the root cap and epidermis; the inner four contribute to the radicle.

The embryo is at first globular, but it soon becomes heart-shaped by a combination of numerous cell divisions and enlargement in two zones of the outer hemisphere. In this manner two cotyledons form. The volume of tissue between the cotyledons is the prospective shoot apex. The characteristic form of the apex is not established until after germination.

As the cotyledons become extended, the embryo bends, because of physical restraints, to conform to the cavity of the embryo sac. From the heart-shaped phase onward, the core of the hypocotyl and the radicle appears as a cylinder of narrow and elongated cells. This is the parent tissue of the vascular system of the seedling. The surrounding tissue contributes the cortex layer of the stem and root.

The embryogenesis of Capsella illustrates only one of several patterns found among flowering plants. Among dicotyledons, the planes of division of the terminal cell, the form of the suspensor, and the contribution made by the basal cell to the embryo all provide evidence used in determining the embryogenetic plan.

Monocotyledons, flowering plants the seeds of which contain only one cotyledon, share with dicotyledons such as Capsella the main features of early embryogenesis, including the possession of a suspensor and, in most cases, a fairly regular progression of cell divisions to the octant stage. Thereafter the symmetrical growth pattern is lost through the development of the single cotyledon. In the lily family (Liliaceae), generally accepted as a primitive family of monocotyledons, the cotyledon is derived from an octad of cells arising from the terminal cell. The hypocotyl and stem apex are derived from the proximal cell of a short filament formed by the basal cell, and the root comes from the pair of cells next to it. The suspensor forms from the distal cell or cells of the filament. In the more-advanced families of monocotyledons, including the grasses (Poaceae) and orchids (Orchidaceae), embryogenesis is much less regular. The grass embryo possesses structures that do not occur in any other flowering plants—namely, the scutellum, an organ concerned with the nutrition of the seedling, and the coleoptile and coleorhiza, protective sheaths of the young shoot and the radicle. The scutellum arises from octant cells, which also contribute to the cotyledon. The basal cell forms part of the coleoptile and also gives rise to the shoot apex and the tissues of the root and coleorhiza. The embryo is asymmetrical, with the shoot apex lying on one side in a notch, ensheathed by the coleoptile.

In marked contrast, embryogenesis of the orchids is simpler. Except when a suspensor is formed, early cleavages follow no well-defined plan, and the product is an ovoid mass of tissue called the proembryo. No cotyledon, stem apex, or root apex is organized in this early period; these organs do not appear until after germination has occurred.

Nutritional Dependence of the Embryo

During their early growth, the embryos of all vascular plants exist as virtual parasites depending for nutrition on either the gametophyte or the previous sporophyte generation through the agency of the gametophyte or, in the special case of the angiosperms, upon an initially triploid tissue, the endosperm, which is itself nourished by the parent sporophyte.

The early nutrition of the sporophyte in ferns, horsetails, and club mosses such as Lycopodium is clearly provided by the gametophyte. In these groups the young sporophyte produces a multicellular structure, the foot, which remains embedded in the tissues of the gametophyte throughout early development withdrawing nutrients. Ultimately, both shoot and root of the sporophyte grow out from the gametophyte, but, even after the first leaf has begun to photosynthesize and thus to produce its own food, the gametophyte may persist.

In Selaginella, the gametophytes are sexually distinct. The female gametophyte develops within the wall of the megaspore. The archegonia are exposed after the megaspore wall splits, but the gametophyte never escapes completely. After fertilization, the zygote cleaves, and the outer cell produces a long suspensor that pushes the embryo deeply into the tissues of the gametophyte. A foot is then formed, as in Lycopodium, and further development of the embryo continues at the expense of reserves transferred to the megaspore from the preceding sporophyte generation.

There are superficial similarities between the nutritional history of the embryo in gymnosperms and in Selaginella, for, in each, the female gametophyte, dependent upon reserves derived from the sporophyte, acts as an intermediary between one sporophyte generation and the next.

In the pines, the female gametophyte develops within the tissues of the nucellus and acquires abundant food reserves. The proembryo forms after a period of free-nuclear division in the zygote, and the tier of cells above the basal four then elongate to form a suspensor, which pushes the embryonic group deep into the gametophyte. Secondary suspensor cells may form from the basal tier to continue the process. During embryogenesis, the gametophyte continues to grow and to accumulate food materials, which are transferred to the embryo or remain as reserves in the seed.

The female gametophyte of angiosperms never acquires copious reserves, although starch is frequently present in the central cell and sometimes in the egg itself. The unique feature, here, is that the embryo is nutritionally dependent upon the endosperm, a tissue that, in the genetically sense, constitutes a third organism—neither gametophyte nor sporophyte. Furthermore, as a tissue the endosperm manifests several other special characteristics. The nuclei have three chromosome sets and, therefore, three times the deoxyribonucleic acid (DNA) of haploid cells. As nuclear division ends, the amount of DNA per nucleus increases still further, a condition comparable with that in various plant- and animal-gland nuclei, presumably connected with the nutritional function of the endosperm. Nuclear division takes place at first without cell-wall formation so that a coenocyte is produced; later, partitioning of the cytoplasm results in a cellular tissue.

The reserves accumulated in the endosperm include carbohydrates (especially starch), lipids, and proteins. As reserves accumulate, the nuclei of the endosperm cells may undergo deformation and degeneration. In many plants the growing embryo consumes the endosperm before seed maturation; in others, the tissue persists in the seed, providing a reserve for the developing seedling after germination. Endosperm is not formed in certain angiosperms. In such cases the embryo depends on the transfer of nutrients directly from the sporophyte.

Tissues other than the endosperm may become specialized for the early nutrition of the embryo. The antipodal cells of the female gametophyte sometimes acquire glandular properties, as may cells of the nucellus surrounding the embryo sac. In some species the embryo itself develops a suspensor that penetrates the tissues of the parent sporophyte and acts as an absorbing organ.

Dormancy of the Embryo

Among the lower vascular plants (club mosses, horsetails, and ferns), the principal agent of dispersal is the haploid spore and not, as in gymnosperms and angiosperms, the seed, the ripened ovule containing a dormant embryo. Since the embryo of lower pteropsids is not involved in dispersal, it does not usually undergo any marked period of dormancy after the differentiation of the primary organs. Development instead proceeds continuously through dependence upon the gametophyte until the young sporophyte is established as a physiologically independent plant. The embryos of gymnosperms and angiosperms pass into a state of dormancy soon after the differentiation of the primary organs and the sporophyte is dispersed in a seed.

In the period leading up to dormancy, several changes occur in the embryo. The accumulation of reserves in the cotyledons or elsewhere ceases respiratory rate declines rapidly, and cell division, with associated protein and nucleic-acid synthesis, stops. Correlated with these events are cellular changes typical of tissues with low metabolic activity. Especially obvious are the general dehydration of the cells that constitute the seed and the thickening of the cell walls of the ovule to form the seed coat (testa). The product is a structure in which the embryo is protected from temperature extremes by its state of desiccation and is often guarded from further drying and from mechanical or biological degradation by the seed coats. The seed coat often contributes to the maintenance of dormancy by physically impeding the passage of water and gases to and from the embryo, by chemically inhibiting germination, and by mechanically restricting the growth of the embryo.

Germination and Early Growth

Dormancy is brief for some seeds, for example those of certain short-lived annual plants. After dispersal and under appropriate environmental conditions, such as suitable temperature and access to water and oxygen, the seed germinates, and the embryo resumes growth.

Figure shows Monocotyledon (internal structures of a corn seed with stages of germination). Nutrients are stored in the cotyledon and endosperm tissue. The radicle and hypocotyl (region between the cotyledon and radicle) give rise to the roots. The epicotyl (region above the cotyledon) gives rise to the stem and leaves and is covered by a protective sheath (coleoptile). (Bottom) Dicotyledon (internal structures of a bean seed with stages of germination). All nutrients are stored in the enlarged cotyledons. The radicle gives rise to the roots, the hypocotyl to the lower stem, and the epicotyl to the leaves and upper stem.

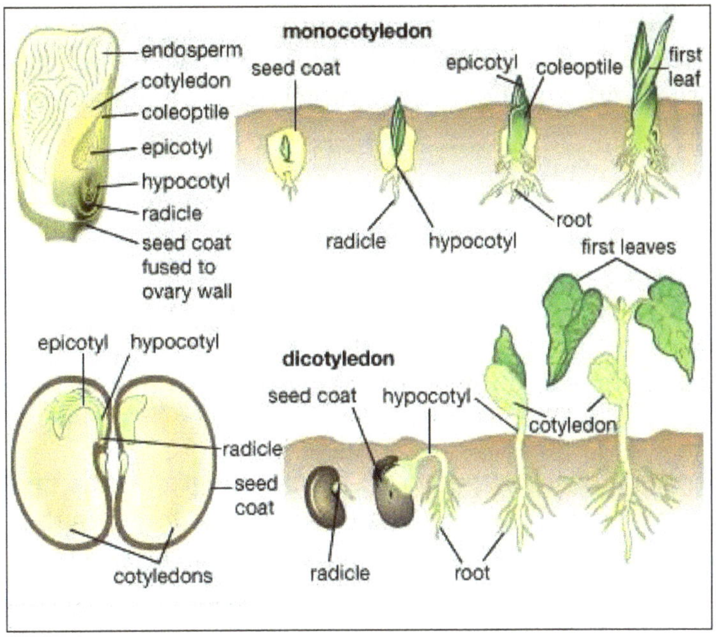

The Breaking of Dormancy

The seeds of many species do not germinate immediately after exposure to conditions generally favourable for plant growth but require a "breaking" of dormancy, which may be associated with change in the seed coats or with the state of the embryo itself. Commonly the embryo has no innate dormancy and will develop after the seed coat is removed or sufficiently damaged to allow water to enter. Germination in such cases depends upon rotting or abrasion of the seed coat in the soil. Inhibitors of germination must be either leached away by water or the tissues containing them destroyed before germination can occur. Mechanical restriction of the growth of the embryo is common only in species that have thick, tough seed coats. Germination then depends upon weakening of the coat by abrasion or decomposition.

In many seeds the embryo cannot germinate even under suitable conditions until a certain period of time has lapsed. The time may be required for continued embryonic development in the seed or for some necessary finishing process—"after ripening"—the nature of which remains obscure.

The seeds of many plants that endure cold winters will not germinate unless they experience a period of low temperature, usually somewhat above freezing. Otherwise germination fails or is much delayed, with the early growth of the seedling often abnormal. (This response of seeds to chilling has a parallel in the temperature control of dormancy in buds.) In some species, germination is promoted by exposure to light of appropriate wavelengths; in others, light inhibits germination. For the seeds of certain plants, germination is promoted by red light and inhibited by light of longer wavelength, in the "far red" range of the spectrum. The precise significance of this response is as yet unknown, but it may be a means of adjusting germination time to the season of the year, or of detecting the depth of the seed in the soil. Light sensitivity and temperature requirements often interact, the light requirement being entirely lost at certain temperatures.

In the process of germination, water is absorbed by the embryo, which results in the rehydration and expansion of the cells. Shortly after the beginning of water uptake, or imbibition, the rate of respiration increases, and various metabolic processes, suspended or much reduced during dormancy, resume. These events are associated with structural changes in the organelles (membranous bodies concerned with metabolism), in the cells of the embryo.

The Emergence of the Seedling

Active growth in the embryo, other than swelling resulting from imbibition, usually begins with the emergence of the primary root from the seed, although in some species (e.g., the coconut) the shoot emerges first. Early growth is dependent mainly upon cell expansion, but, within a short time, cell division begins in the radicle and young shoot; thereafter, growth and further organ formation (organogenesis) are based upon the usual combination of increase in cell number and enlargement of individual cells.

Until it becomes nutritionally self-supporting, the seedling depends upon reserves provided by the parent sporophyte. In angiosperms these reserves are found in the endosperm, residual tissues of the ovule, or in the body of the embryo, usually in the cotyledons. In gymnosperms, food materials are contained mainly in the female gametophyte. Since reserve materials are partly in insoluble form—as starch grains, protein granules, lipid droplets, and the like—much of the early metabolism of the seedling is concerned with mobilizing these materials and delivering, or translocating, the products to active areas. Reserves outside the embryo are digested by enzymes secreted by the embryo and, in some instances, also by special cells of the endosperm.

In some seeds (e.g., castor beans) absorption of nutrients from reserves is through the cotyledons, which later expand in the light to become the first organs active in photosynthesis. When the reserves are stored in the cotyledons themselves, these organs may shrink after germination and die or develop chlorophyll and become photosynthetic.

Environmental factors play an important part not only in determining the orientation of the seedling during its establishment as a rooted plant but also in controlling some aspects of its development. The response of the seedling to gravity is important. The radicle, which normally grows downward into the soil, is said to be positively geotropic. The young shoot, or plumule, is said to be negatively geotropic, because it moves away from the soil; it rises by the extension of either the hypocotyl, the region between the radicle and the cotyledons, or the epicotyl, the segment above the level of the cotyledons. If the hypocotyl is extended, the cotyledons are carried out of the soil, but, if the epicotyl elongates, the cotyledons remain in the soil.

Light affects both the orientation of the seedling and its form. When a seed germinates below the soil surface, the plumule may emerge bent over, thus protecting its delicate tip, only to straighten out when exposed to light (the curvature is retained if the shoot emerges into darkness). Correspondingly, the young leaves of the plumule in such plants as the bean do not expand and become green except after exposure to light. These adaptative responses are known to be governed by reactions in which the light-sensitive pigment phytochrome plays a part. In most seedlings, the shoot shows a strong attraction to light, or a positive phototropism, which is most evident when the source of light is from one direction. Combined with the response to gravity, this positive phototropism maximizes the likelihood that the aerial parts of the plant will reach the environment most favourable for photosynthesis.

Later Development: The Sporophyte Plant Body

Continuation of Organ Formation

General Sherman tree: The General Sherman tree, the world's largest (in bulk) giant sequoia (Sequoiadendron giganteum), in Sequoia National Park, California.

Although it is convenient to refer to the early development of the plant sporophyte from the fertilized egg as embryogenesis, the process is never actually concluded as it is in the higher animals. In vascular plants, organ formation (organogenesis) is not confined to early life, and the processes of shoot, root, and leaf formation that occur first in the embryo are repeated, albeit in modified form, throughout the life of the plant. The life span may be short and determinate, as in annual plants such as the cereals, or long, lasting for many years—indeed potentially indefinitely, except for limitations imposed by the environment and accidents—as in trees. The protracted growth of perennials, or plants that resume growth each growing season, tends to lead to increase in size, but bulk is not necessarily directly correlated with age, because individual leaves, flowers, and even whole limbs continuously die and are shed. Some long-lived plants, however, do reach a point at which losses of body mass balance the increase resulting from continued growth and organ formation.

The Activity of Meristems

Characteristically, vascular plants grow and develop through the activity of organ-forming regions, the growing points. The mechanical support and additional conductive pathways needed by increased bulk are provided by the enlargement of the older parts of the shoot and root axes. New cells are added through the activity of special tissues called meristems, the cells of which are small, intensely active metabolically and densely packed with organelles and membranes, but usually lacking the fluid-filled sacs called vacuoles. Meristems may be classified according to their location in the plant and their special functions. One important distinction is between persistent meristems, typified by those of the growing points, and meristems with a limited life, those associated with organs, such as the leaf, of determinate growth. The regions of rapid cell division at the tips (apices) of the stem and the root are terminal meristems. In the stem apex, the uppermost part is the promeristem, below which is a zone of transversely oriented early cell walls, the file, or rib, meristem. The procambium is a meristematic tissue concerned with providing the primary tissues of the vascular system; the cambium proper is the continuous cylinder of meristematic cells responsible for producing the new vascular tissues in mature stems and roots. The cork cambium, or phellogen, produces the protective outer layers of the bark.

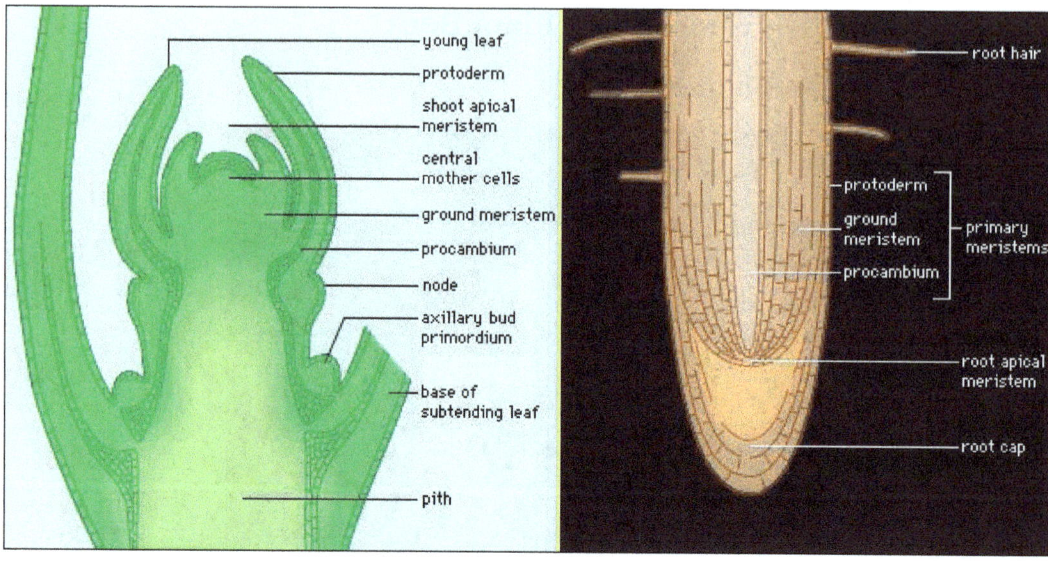

Root and shoot apical meristems.

Figure shows the shoot apical meristem of Hypericum uralum (left) appears at the topmost aspect of the stem. Immediately behind the apical meristem are three regions of primary meristematic tissues. The root apical meristem (right) appears immediately behind the protective root cap. Three primary meristems are clearly visible just behind the apical meristem.

Among meristems of limited existence is the marginal, or plate, meristem responsible for the increase in surface area of a leaf; it contributes new cells mainly in one plane. Another type of meristem of limited life is called intercalary; it is responsible for the extension of some stems (as in the grasses) by the addition of new tissues remote from the growing points.

The number of dividing cells in persistent meristems remains roughly constant, with one of the daughter cells of each division remaining meristematic and the other differentiating as a component of a developing organ. The geometrical arrangements in the particular organ determine the way in which this occurs, but in general the consequence is that the meristem is continuously moving away from the maturing tissue as growth continues. It remains, therefore, a localized zone of specialized tissue, never becoming diluted by the interposition of expanding or differentiating cells. In organs such as leaves, flowers, and fruits, in which the growth is determinate, the divisions of meristematic cells become more widely scattered, and the frequency progressively falls as the proportion of the daughter cells that differentiate increases. Ultimately, at maturity, no localized meristem remains.

The Contribution of Cells and Tissues

The two major factors determining the forms of plant tissues and organs are the orientation of the planes of cell division and the shapes assumed by the cells as they enlarge. Clearly, if the division planes in a cell mass are randomly oriented and individual cells expand uniformly, the tissue will enlarge as a sphere. On the other hand, if cell division planes are oriented regularly or the expansion of individual cells is directional, the tissue can assume any of a number of shapes. In a stem, for example, the cell division planes of the promeristem are oriented at various angles to the stem axis, so that new cells produced contribute to both width and length. Below this region, in the rib meristem, the proportion of divisions with the cell plate at right angles to the axis increases, so that the cells tend to be oriented in files. The cells in these files expand vertically more than they do horizontally, and, accordingly, the stem develops as a cylinder.

The factors that control the orientation of cell division planes in meristems are largely unknown. Cell interactions, however, are presumed to coordinate the distribution and orientation of the divisions. In each cell microtubules in the cytoplasm help to orient the nucleus before it divides. Then, at the time of the division, other microtubules arranged in a spindle-shaped figure (the mitotic spindle) are involved in separating the daughter chromosomes and moving them to opposite ends of the parent cell. Thereafter, the residual part of the spindle helps to locate the plate that separates the two daughter cells. Microtubules are also concerned in determining the direction of growth in expanding cells, since they appear to influence the construction of the cell wall by controlling the way cellulose is laid down in it.

Although change in shape is a form of cell differentiation, the term in the more general sense refers to a change in function, usually accompanied by specialization and the loss of the capacity for further division. Biochemical differentiation often involves a change in the character of the cell organelles—as when a generalized potential pigment body (proplastid) matures as a chloroplast, a chlorophyll-containing plastid. But it may also involve structural changes at a subcellular level, as when organelles change their character in cells engaged in intense metabolic activity.

The differentiation of plant cells for the movement of materials and the provision of mechanical support or protection invariably depends upon modification of the walls. This usually entails the accretion of new kinds of wall materials, such as lignin in woody tissue and cutin and suberin in epidermal tissues and cork. The accompanying structural changes must be controlled, for the wall materials are not applied at random but according to a pattern appropriate to the particular cell or tissue. The development of patterns during cell-wall growth depends not only on the cytoplasmic microtubules, as in the construction of the cells that will give rise to the water-conducting vessels (xylem elements), but also on cytoplasmic membranes, as in the formation of sievelike end walls (sieve plates) in the cells that will give rise to food-conducting vessels (phloem elements).

The differentiation of xylem culminates in the death of the participating cells, and the vessels are formed of chains of empty walls. This is an example of "programmed death," not an uncommon phenomenon in plant and animal development.

The Shoot System and its Derivatives

The Shoot Tip

The gametophytes of mosses and liverworts and the sporophytes of many higher plants have a shoot, or early stem, with a single cell at its tip, or apex, from which all the tissues of the stem arise. This apical cell is usually four-sided (tetrahedral), with three faces directed downward, and the fourth capping the apex. Daughter cells are continually cut off sequentially from the three inner faces, the apical cell preserving its tetrahedral shape. In cell lineages derived from the daughter cells, the division planes may remain oriented in a more or less regular manner, so that, for some distance below the apex, the three sectors can be recognized in the stem. This basic pattern occurs in the arrangement of the "leaves" of some mosses, which lie in three ranks. In many plants, however, division planes in the lower part of the apex show no particular correlation with the planes of cleavage of the apical cell, and the lateral appendages do not reflect any three-part arrangement.

Gymnosperm and angiosperm apices do not possess apical cells. The generative role is discharged by an ill-defined zone of tissue called the promeristem. Regularities may appear in the distribution of division planes only in the extreme tip region. Over the outer part of the apex, the cells often appear to lie in one to three layers, which constitute the tunica. Enclosed by the tunica lies a core of cells that exhibits no distinct layering; this zone is the corpus. The layers of the tunica normally contribute to the surface layers of the plant, and the corpus provides the deeper lying tissues.

The tunica–corpus analysis emphasizes the orientation of division planes, but apices can be examined from other points of view—the sizes of cells, the degree of vacuolation, and the concentration of various cell constituents, especially ribonucleic acid (RNA), vary through the apex and this sometimes results in more or less distinctive zones. Both gymnosperm and angiosperm apices have been classified on the bases of such zonal patterns, but the validity of this approach, as well as its usefulness for understanding the function of the apex as a morphogenetic centre, has been questioned.

Since 1950, a theory of angiosperm apical zonation developed by French and Belgian botanists has been gaining support. This theory proposes that the central region of the apical dome constitutes a mass of cells with relatively low division rates, the méristème d'attente, or "waiting meristem." Surrounding this region is an annular zone of cells with higher division rates, the anneau initial, or "initiating ring." Features other than division rates characterize these zones: RNA and protein content are lower in the méristème d'attente than in the anneau initial, and the nucleoli are smaller. In longitudinal section, the differences contribute to the patterns distinguishable in apices, some of which have been used as bases for structural classification. The main contention of the Franco-Belgian school, however, is that the zonation represents a functional difference. The méristème d'attente is regarded as a region mainly concerned with controlling the geometry of the apex. The cells have a restricted metabolism concerned primarily in maintaining a low rate of increase in cell number, and they themselves, as well as their immediate derivatives, take no part in organogenesis or associated differentiation. The anneau initial, by contrast, is that part of the apex that produces the beginnings, or primordia, of lateral organs. Not only is the division rating higher, but the tissue as a whole is involved in metabolic syntheses that precede morphogenesis.

One difficulty in investigating the stem apex arises from the uncertainty about which aspects are important for the overall function: division planes, division frequency, metabolic patterns, or some combination of these. Still another complication results because the apex is in a state of constant change during the growth of the plant. A long-term developmental trend begins after the definition of the growing point in early embryogenesis and continues thereafter through juvenility and the period of vegetative growth into the reproductive phase. Superimposed on this trend is a cyclical change reflecting the periodic generation of the primordia of leaves and lateral shoots in the region immediately under the apex.

The Production of Leaves

Leaves originate on the flanks of the shoot apex. A local concentration of cell divisions marks the very beginning of a leaf; these cells then enlarge so as to form a nipple-shaped structure called the leaf buttress. The cells of the leaf buttress may be derived from the tunica alone or from both the tunica and the corpus.

In the early growth of the leaf primordium, new cells are contributed mainly by meristematic activity at the pole directed away from the stem, so that the buttress extends in length. The subsequent distribution of growth varies among the different groups of vascular plants according to the shape of the mature leaf. In considering the angiosperms, a broad-leaved dicotyledon (e.g., tobacco) and a narrow-leaved monocotyledon (e.g., maize [corn]) will serve as examples.

Apical growth dominates in the tobacco-leaf primordium until a height of about 0.5 millimetre (0.02 inch) is reached. Thereafter, the buttress becomes more and more flattened in the transverse plane by laterally oriented cell divisions and further expansion growth on either side. The dividing zones are the marginal meristems, through the activity of which the leaf gains its laminate form. In each meristem the outer file of cells, or marginal initials, contributes the epidermal layers by continued division. The cells below, the submarginal initials, provide the tissue of the inner part of the leaf. Usually a certain number of cell layers are defined in the mesophyll (the parenchyma between the epidermal layers of a foliage leaf). Cell division is not limited to the region of the marginal meristems but continues throughout the leaf in each of the layers, always in the same plane, until the final cell number is approached. The rate then declines, ceasing in the different layers at different times. Divisions usually end first in the epidermis, then in the lower mesophyll layers of a leaf such as that of tobacco, and last in the main photosynthetic tissue, the palisade layer, just beneath the upper epidermis.

The vascular pattern in a tobacco leaf is determined early in the development of the vessel primordium. A procambial strand is formed by the elongation of narrow axial cells, and this extends both toward the base and toward the apex, eventually linking with the procambium of the stem. When the marginal meristems become active, the lateral veins of the leaf are initiated first, followed by the third and later order branchings that give the characteristic network of veins in the mature leaf.

Although the differentiation of the cells of the vascular system begins at the base, the epidermal and mesophyll cells mature from the tip inward toward the stem. The palisade cells elongate in a plane at right angles to the epidermis; those of the lower mesophyll expand irregularly to give lobed forms. The cells of the epidermis, shaped like irregular paving stones, continue to expand in the plane of the leaf after growth ceases in the mesophyll, so that the cells of the internal tissues are pulled apart to form the system of air spaces found in the mature leaf.

Dicotyledonous leaves are folded in various ways in the bud, the patterns being determined by differential growth in the tissues of the upper and lower surfaces (laminae) of the young leaves. Differential growth may cause the lamina either to roll or fold toward the leaf midrib or to fold near lateral veins, thus pleating the lamina. The folds in the bud are, of course, eliminated during the final phase of leaf expansion.

In the development of the maize leaf, the primordium arises first as a prominence some distance below the apical dome. The zone of division and growth extends laterally around the apex so that a complete collar forms; then the margins overlap. Meanwhile the original tip zone continues to elongate, eventually surpassing the stem apex. Tip growth declines thereafter, and further increase in cell number results from meristematic activity at the base. The early development of the vascular system is unlike that in dicotyledons, for several parallel procambial strands, rather than a single midrib, are initiated. The first of these grow toward the apex, but, as tip growth ceases, procambial strands form above and extend toward the base, passing through the node, or point of insertion of the leaf primordium, and into the stem below. As the leaf extends in length, the tissues begin to mature first at the tip, and a wave of differentiation passes down toward the base, where cell division and extension growth may continue long after the tip of the leaf is mature. Protection for this immature and succulent tissue of the leaf base is afforded by the sheaths of older leaves surrounding it.

These examples illustrate the principles involved in leaf development, but there are many deviations associated with variations in leaf form. Lobing and toothing result from the persistence of cell division and growth in particular stretches of the margin after growth ceases in between. Carried to the extreme, this localized growth gives the feather-like pinnate leaf. Many monocotyledons form cylindrical leaves as a result of a fusion of the margins of the primordium after it has encircled the stem.

Branching of the Shoot

The shoots of most vascular plants branch according to a consistent plan, with each new axis arising in the angle between a leaf and a stem—that is, in a leaf axil. In some plants, buds may also form from the older parts of shoot or root remote from the main apices; these buds, termed adventitious, do not conform to the general plan.

A lateral shoot apex is initiated on the flanks of the main apex but at some distance below the point of emergence of the youngest leaf primordium. As in the origin of a leaf, generally the outer cell layers contribute to the surface tissues of the new apex by maintaining a consistent pattern of divisions. In some species a tunica of more than one cell layer quickly forms, so that the new apex appears as a miniature version of the main one; alternatively, the differentiation may not become apparent until the new primordium has attained considerable bulk. In all cases, the new apex must reach a minimal volume before it in turn can begin to form its own lateral primordia and to organize true axillary buds. As this volume is attained, zonation appears. As in the main apex, the formation of new primordia is associated with the annular zone.

From this point on, the development of the lateral shoot is the same as that of the main shoot, except that growth may not be as rapid because the main apex, or leading bud, dominates and absorbs much of the available nutrient. The early growth of the axillary bud proceeds quite vigorously until a certain number of leaf primordia has been formed; then apical activity slows. Cell division gradually stops, and with it the associated syntheses; thus there is no increase in the DNA of the nuclei of the meristem after the last division. The bud, in effect, passes into a state of dormancy, even though the external conditions for growth are propitious. This phenomenon is known as correlative bud inhibition, since it is determined by the activity of the leading bud of the shoot. If the leading bud is removed, the inhibited lateral buds resume growth, and with it the associated syntheses.

Vascular Development

Cell division planes in the zone just below the apex of the shoot tend to be oriented so that vertical files of cells are formed. This is more evident in the central core than in the surrounding cortical region, for the pattern is not disturbed by the insertion of lateral members. The first signs of the differentiation of the vascular system appear some distance below the apex, in a zone of tissue distinguishable by the smaller cross-sectional area of individual cells. These cells, forming the procambium zone, arise by divisions oriented at right angles to the axis and may form a complete cylinder; generally, however, interruptions occur, the segments being related to the uppermost leaf primordia. In a dicotyledon such as tobacco, the cylinder at its highest level consists of strands running upward toward the points of insertion of the primordia. Thus, as the site of each primordium is determined, a strand forming in the adjacent region of the stem will contribute to the cylinder, but at a higher level than the preceding strand. The link with the earlier formed procambium is not simple, however. The strand passing upward toward a leaf primordium usually is composed of branches arising from strands that enter the two nearest older leaves below it. Because it in turn will contribute a branch for the next leaf, the cylinder is really a hollow network, the "gaps," or leaf traces, marking the points of departure of the leaf veins.

During subsequent development, the strands, or vascular bundles, increase in thickness by further cell divisions, and connections form with the vascular systems of axillary buds. The cells differentiate to give the characteristic tissues of the vascular system: phloem vessels (conducting tissue), phloem parenchyma (packing tissue), and phloem fibres (supporting tissue) toward the outside and xylem vessels (woody conducting tissue) toward the inside. The differentiation occurs in an upward direction, so that the maturation of the vascular tissues follows at a more or less constant distance behind the apex.

Although details differ, the above account of the origin of the primary vascular system is broadly applicable to gymnosperms and many ferns. Vascular development differs somewhat in certain flowering plants. In many monocotyledons, such as maize, the several vascular strands that pass down from each leaf primordium into the stem do not contribute to a single cylinder but are scattered in the ground tissue, or parenchyma, of the stem. Lateral interconnections form principally at the nodes.

Increase in stem diameter is accomplished in the older stems of dicotyledons by the activity of the cambium, which produces secondary vascular tissue. This meristem, a relic of the procambium, is composed of thin-walled cells, is flattened in the radial plane, and persists between primary vascular tissue, the differentiated outer phloem, and the inner xylem. When secondary thickening begins, the parenchymatous cells between the vascular bundles also resume division, ultimately forming a cambium cylinder. The cells of the cambium divide, producing initial phloem cells toward the outside and initial xylem cells within. Files of cells are also cut off among the initial phloem and xylem cells that remain parenchymatous and are called phloem and xylem rays.

As in the apical meristem, the number of dividing cells in the cambium remains constant, except that more cells occasionally are added by divisions in the radial plane so that the girth of the cambial cylinder expands in pace with the growth of the xylem within. The addition of new phloem toward the outside compresses the primary phloem and the cortical tissues in a radial direction while stretching them tangentially. These primary tissues do not persist, however. As the girth of the stem increases, the epidermis is disrupted, and the outer layers of the cortex become meristematic, giving rise to the cork cambium, which generates cork cells on the outside. The cork layer, or bark, then takes over the protective function of the epidermis.

The Root System and its Derivatives

The Root Tip

Plants that have a single apical cell in the shoot also have a single apical cell in the root. The cell is again tetrahedral, but sometimes daughter cells are cut off from all four faces, with the face directed away from the axis producing the cells of the root cap. The cells derived from the other faces continue to divide mostly by forming transverse walls, but occasionally also in the longitudinal plane. In this way vertical columns of cells form—tending, because of their mode of origin, to be disposed in three sectors.

In the roots of gymnosperms, angiosperms, and some lower plants, there is no single apical cell. Again, as with the shoot, such root apices can be analyzed in different ways. Perhaps the most useful approach is based upon tracing the sources of the main tissues in the apical region. Such an analysis has led to the histogen theory, which proposes that the three principal tissues of the root—vascular cylinder, cortex, and epidermis—originate from three groups of initial cells, or histogens, in the apical meristem—plerome, periblem, and dermatogen respectively. A fourth histogen, the calyptrogen, produces the root cap. The histogens have been thought to lie in linear order in the apex, with the initial cells of the vascular system toward the older part of the root, and those of the cap toward the tip.

The histogen theory is difficult to apply to some types of roots, and there has been uncertainty about the numbers of histogens. The discovery of the "quiescent centre" in the root apex has clarified many features, however. The quiescent centre is a group of cells, up to 1,000 in number, in the form of a hemisphere, with the flat face toward the root tip; it lays at the centre of the meristem, in much the same position, in fact, as the tetrahedral apical cell in certain lower plants. The cells of the quiescent centre are unusual in that their division rate is lower than that in the surrounding meristem. The cells of the centre have other distinctive features as well, notably a lower rate of protein synthesis than that of neighbouring cells.

The quiescent centre is surrounded by actively dividing cells of the promeristem that are the initial cells of the various tissues of the root. Those abutting the flat, tip-directed face contribute to the root cap; those above the quiescent centre are distributed in a cup shape. The cells in the centre of the cup produce the procambium and so, ultimately, give rise to the vascular cylinder. The annular zone of cells surrounding this central group forms the initials of the cortex; surrounding this, in turn, a ring of initial cells forms the protoderm, the layer corresponding to the epidermis at this level of the root.

The quiescent centre, a constant feature of the root tip, is apparently generally present in angiosperm and probably also in gymnosperms. The quiescent centre probably plays a role comparable with that of the apical cell in some lower plant roots, maintaining the geometry of the system. It has also been suggested that it may be concerned with the synthesis of growth hormones, although no direct evidence exists. When roots are damaged mechanically or by radiation, the cells of the centre can resume a rapid division rate, and they then participate in regeneration.

The zone of cell division extends some distance along the length of the root above the tip region. Although the girth may increase by longitudinal divisions and the widening of the daughter cells, most divisions occur in the transverse plane resulting in the formation of longitudinal files of cells.

In longitudinal section, the tissue zones become progressively better defined away from the tip. An internal protective band, the endodermis, becomes conspicuous as a single sheath of cells surrounding the procambium. The phloem procambium, recognizable by its narrow cells, begins to differentiate in the lower part of the region of elongation. The xylem also becomes distinct, the thickenings appearing first in the upper part of the extension zone. Differentiation keeps pace with the advance of the root tip as new cells are added in the promeristem. When xylem occupies the core, there is no pith as in the shoot, but the cells of the outermost layers of the vascular cylinder remain undifferentiated, forming the pericycle, a tissue important in the formation of lateral roots. Within the bounds of the pericycle, the xylem is star-shaped in section, with the first-formed xylem elements (protoxylem) occupying the ridges. The phloem lies in the intervening grooves. Outside of the endodermis, the cortical cells elongate but remain thin-walled. Above the level of the root-cap sheath, the epidermis forms the outer layer of the root, and, beyond the extension zone, its cells begin to develop root hairs.

A more complete account can be given for the mechanics of development of the root apex than for that of the stem, mainly because of its greater simplicity. An important difference lies in the absence of a mechanism for the cyclical production of lateral organs at the apex itself.

Branching of the Root

The branching of the root takes place in the older parts and does not directly involve the apical meristem. The tissues concerned are the endodermis and the layer immediately beneath it, the pericycle. The endodermis participates in root branching in certain lower plants with apical cells. A cell of this layer enlarges and forms a tetrahedral cell, which becomes the new apical cell; by further divisions a hemispherical volume of tissue forms around it—the whole constituting a new apex.

In many other plants, including gymnosperms and angiosperms, the lateral roots develop from the pericycle. Cells in this layer enlarge and begin to divide until a dome of tissue develops. Called the incipient apex, the dome pushes out the surrounding endodermis, which may itself resume divisions, its daughter cells enlarging to create a sheath around the new root tip. During further growth, the dome assumes an organization like that of the primary root apex. At first, all cells are meristematic; then, while the primordium is still small, cells in the central zone cease DNA synthesis, and this zone becomes the new quiescent centre. Beyond it, the root cap is produced, and, at the base, initial cells begin to develop the cell files that become the vascular cylinder, cortex, and epidermis. The vascular tissues differentiate from the base outward, and link eventually with xylem and phloem of the parent root. All this development occurs before the tip of the new root emerges from the tissues of the parent root. The growth of the new tip into the cortex first pushes out the endodermal sheath, if one is present, and then bursts it. The cortical cells are themselves crushed and probably resorbed as the root grows on, until finally the tip breaks through the epidermis.

In most roots, new laterals are initiated in the pericycle opposite to the protoxylem ridges. They tend accordingly to form vertical ranks along the length of the root, reflecting the number of bands of protoxylem. Although lateral roots arise in quite a different way from leaves and axillary shoots at the stem apex, there are certain common features. Pericyclic cells about to produce a root primordium synthesize ribonucleic acid, in anticipation of the period of growth and morphogenesis that will result in a new apex. The same behaviour is seen in the cells of the annular zone, from which leaf primordia arise at the stem apex, and also in the axillary zones at a slightly lower level, from which new stem apices develop.

Later Growth

In the secondary growth of the root, cell division in the primary xylem produces a cambium, which abuts the pericycle over the protoxylem ridges and passes between the phloem strands and the xylem in the grooves. Activity of the root cambium is comparable with that of the stem cambium; phloem elements are cut off outward, and xylem elements are cut off within. With continued growth in thickness, the star-shaped figure of the primary xylem is lost, and the cambium eventually forms a cylindrical sheath. Again, as in the stem, the protective function of the epidermis is ultimately taken over by cork layers produced by a cork cambium in the outer cortex.

Correlations in Plant Development

Coordination of Shoot and Root Development

Although the structural organization of the vascular plant is comparatively loose, development of the various parts is well coordinated. Control is dependent upon the movement of chemical substances, including both nutrients and hormones.

An example of correlation is the growth of shoot and root. The enlargement of aerial parts is accompanied by increased demands for water, minerals, and mechanical support that are met by coordinated growth of the root system. Several factors apparently are concerned with control, because shoot and root affect each other reciprocally. The root depends on the shoot for organic nutrients, just as the shoot depends on the root for water and inorganic nutrients and the flow of ordinary nutrients must, therefore, play some part. More specific control, however, may be provided by the supply of nutrients required in very small amounts. The root depends on the shoot for certain vitamins, and variation in the supply, reflecting the metabolic state of the aerial parts, may also influence root growth. In addition, hormonal factors affecting cell division pass upward from the root into the stem; although the exact role of the hormones has not yet been established with certainty, they may provide one way by which the root system can influence the activity of the shoot apex.

The control of secondary thickening is another important example of growth correlation. As the size of the shoot system increases, the need for both greater mechanical support and increased transport of water, minerals, and manufactured food is met by an increase in stem girth through the activity of the vascular cambium. Generally, the cambium of trees in temperate zones is most active in the spring, when buds open and shoots extend, creating a demand for nutrients. Cell division begins near the bud in each shoot and then spreads away from it. The terminal bud stimulates the cambium to divide rapidly through the action of two groups of plant hormones: auxins and gibberellins.

The inhibition of lateral buds, another example of correlated growth response, illustrates a reaction opposite to that occurring in the control of cambial activity. Lateral buds are inhibited in general because axillary shoots grow more slowly or not at all, while the terminal bud is active. This so-called apical dominance is responsible for the characteristic single trunk growth seen in many conifers and in herbaceous plants such as the hollyhock. Weaker dominance results in a bushy growth form with repeated branching. The fact that lateral, or axillary, buds become more active when the terminal bud is removed suggests that hormonal control is involved.

The flow of auxin from the shoot tip is, in part, responsible for inhibiting axillary buds. The nutritional status of the plant also plays a role, apical dominance being strongest when mineral supply and light are inadequate. Because axillary buds are released from inhibition when treated with cell-division promoting substances (cytokinins), it has been suggested that these substances are also concerned in regulating axillary-bud activity.

Determination of Mature Form

After its establishment as an independent plant, the sporophyte passes through a juvenile period before reaching maturity and becoming reproductive. Juvenility may be brief or, as in the case of trees, may extend over several years. The duration is determined partly by internal factors and partly by environmental controls related to the seasons.

Internal Control of Development

In some ways juvenility is a continuation of developmental trends initiated in the embryo. In many plants, new organs are produced sequentially through early life, each of progressively more mature form. The first leaf of the young fern sporophyte, for example, is small and relatively simple, and the vascular system consists of a few forked strands. As growth proceeds, succeeding new leaves are of increasing complexity, and the shape begins to resemble that typical of the reproductive frond; in addition, vasculation shifts to the mature pattern, often one with a network of veins. Comparable trends occur in flowering plants, in which leaves at successive levels of plant maturity often show a progressive increase in the complexity of lobing or toothing.

Some of the changes associated with the juvenile period can be attributed to the gradual enlargement of the growing point, necessarily small in the embryo; its volume increases progressively with development. This increase in cell number is usually associated with the emergence of a "mature" zonation pattern. The typical internal structure of the shoot apex does not develop until a specific number of leaves form.

Gradual structural change in the growing point, however, does not adequately account for all aspects of juvenility. Sometimes, the transition from juvenile to adult leaf form is not graded but sudden. The juvenile leaves of species of the gymnosperm *Chamaecyparis*, for example, are needlelike and spreading; the adult leaves are scale like and lie close to the stem. Among flowering plants, various species of *Eucalyptus* have juvenile leaves that are ovate and mature leaves that are sickle-shaped.

Such sudden transitions from juvenile to adult form, referred to as phase change, seem to depend not on slow shifts in the apex but on some determinative event or correlated group of events. The two forms are relatively stable and tend to resist change; for example, cultured tissues taken from the juvenile parts of ivy plants maintain a higher rate of cell division, and portions, or cuttings, taken from these parts tend to form roots more readily than those from the adult parts.

The establishment of these relatively stable but not wholly irreversible states is comparable with the determination of shoot and root poles during embryogenesis and, indeed, with the alternation of generations itself. The transmission of differentiated states through cell lineages presumably reflects the action of "switching" devices controlling the expression of different parts of the genetic complement. In this sense, phase change and related phenomena do not differ essentially from those of differentiation and organogenesis in general.

The transition in plants to the reproductive state is an example of a developmental event with some of the characteristics of phase change. Among seed plants, the reproductive structures are transformed shoots—strobili (including cones) of various kinds in the gymnosperms and flowers in angiosperms.

From a developmental point of view, the flower can be regarded as a shoot axis of determinate growth, with the lateral members occupying the sites of leaves differentiating as floral organs—sepals, petals, stamens, and pistils. In the transition to flowering, the stem apex undergoes distinctive changes, the most conspicuous of which is in the shape of the apical region, which is related to the kind of structure to be formed, whether a single flower, as in the tulip, or a cluster of flowers (an inflorescence), as in the lilac. The region of cell division extends over the entire apex, and the RNA content of terminal cells increases. When a single flower forms, lateral primordia emerge at higher and higher levels on the flanks of the apical dome, and the entire apex is absorbed in the process, after which apical growth ceases. When an inflorescence forms, early changes are generally comparable to that for the single flower with one major difference—axillary primordia emerge that either become floral meristems or develop as secondary inflorescence branches. These primordia appear closer to the apex than do those of axillary buds on a vegetative shoot. In grasses, the activation of axillary meristems is the most notable early indication of the passage into flowering.

Environmental Control of Development

The rate of maturation and the timing of the transition to the reproductive phase are sometimes governed by internal controls and thus are relatively insensitive to the environment, provided conditions are generally favourable for growth. Frequently, however, the developmental rate is affected profoundly by recurring cycles in the environment, particularly those of temperature and of day length. In effect, these cycles provide a timetable for the plant, thus adjusting flowering, fruiting, and seed dispersal to the season and increasing the chances for successful propagation.

The control of the developmental rate by temperature is especially evident in many herbaceous plants of temperate climates. These plants, as indicated earlier, often must experience cold, either as seeds or as young plants, before they can begin flower production; otherwise they undergo an excessively long period of leafy, or vegetative, growth. After the cold experience, which can be given artificially, the plant is said to have been vernalized, or brought to the spring condition. Again the response is akin to a determination, because the condition attained is transmitted through subsequent cell divisions. Furthermore, there are indications that vernalization induces a persistent modification in the metabolism of apical cells and their derivatives.

The annual cycle of changing day length obviously provides the best of all "clocks" for the regulation of plant development. The effect of day length (or rather length of continuous darkness) on the transition to flowering is part of the general phenomenon of photoperiodism. Certain plants, called short-day plants, grow vegetatively when the nights are shorter than a critical minimum period (days long); exposure to longer nights (days short), however, accelerates development and brings on early flowering. Conversely, long-day plants develop very slowly toward flowering during daily cycles with longer than a minimum of darkness (days short), and are accelerated by exposure to short nights (days long). Other plants either require days of intermediate length for flowering or respond to a sequence of different photoperiods.

The leaf, rather than the stem apex, is the light-receiving organ in the photoperiodic reaction, although it is at the apex that subsequent developmental changes occur. As in the case of vernalization, photoperiod undoubtedly affects the metabolism of the known plant hormones, and so influences many other developmental responses apart from flowering. The effect of the duration of illumination on the carbohydrate balance of the plant may also be important. Nutritional effects on flowering are well known in many species—certain fruit trees, for example.

Whether or not environmental factors influence the passage into a reproductive state of a plant, the transition must be viewed as part of the general development from juvenility to maturity: in this sense, flowering is not a radical alternative to vegetative growth but its culmination. Yet, entirely new organ types are produced at the flowering apex, presumably under the influence of genes inactive during vegetative growth.

Seasonal Adaptations

Certain plants are perennial and survive from year to year by matching their growth to the progression of the seasons or by suspending growth altogether during unfavourable times, such as winter or a dry season.

In the temperate zone, sometime before winter begins, growth ceases in the shoots of woody plants, resting buds are formed, and deciduous trees lose their leaves. The resting bud consists of a short axis, with the stem apex surrounded by modified unexpanded leaves, which protect the stem, especially from drying. The cells show marked frost resistance, similar to that of the embryo of the seed. Corresponding changes occur in herbaceous plants, in which the preparation for winter may involve the dying back of aerial parts altogether, leaving protected organs at or below the soil surface.

Growth sometimes ceases, even under favourable conditions, as a result of internal changes in the plant. This is true for some trees, which cease growth in midsummer. The passage into winter dormancy, however, is often controlled by the shortening of day length at the end of the growing season; in some plants, decreasing night temperature also plays a part. Most temperate-zone trees cease growth and form resting buds when the day-length falls below a critical minimum.

Photoperiodic control seems to involve the formation of inhibitor compounds. In birches, for example, the leaf perceives the day-length "signal" and transmits inhibitory materials to the apex, thus bringing growth to a stop and inducing the formation of a resting bud. The dormancy hormone, abscisic acid, may be concerned in this response and also in stomata closure in response to water stress and in leaf abscission (the separation of the leaf from the plant); there is a great deal of debate among scientists concerning what role, if any, this hormone plays in abscission. Other signaling hormones, such as indole-3-acetic acid (an auxin) and ethylene, appear to play a more significant role in leaf abscission.

Budbreak in certain trees is controlled by photoperiod, growth resuming in the lengthening days of spring; light-perceptive organs are probably the young leaves inside the bud scales. Sometimes budbreak depends only on temperature increases that occur in spring, as in certain plants of Mediterranean climates.

The resumption of development in buds may result from a change in the balance of growth-inhibiting substances, such as abscisic acid, and growth promoters, notably the gibberellins. Buds can be caused to open prematurely by gibberellin treatment, which, as in the case of vernalization, can sometimes replace a cold experience; moreover, the gibberellin content in the buds of certain woody plants increases during chilling. Other hormones are probably also involved, however, for budbreak in plants such as the grapevine can be promoted by cytokinins, the plant cell-division factors.

An important general feature of adaptive periodicities is that the developmental changes anticipate the conditions for which they will ultimately provide the appropriate physiological or morphological adjustment. The ability of plants to utilize environmental indicators such as temperature and day-length changes is vital for the survival of plants. The production of such adaptive devices is made possible by the state of continuous embryogeny, already stressed as one of the most important characteristics of plant growth.

Senescence in Plants

The growth of the vascular plant depends upon the activity of meristems, which are, in a sense, always embryonic. Continued indefinitely, this mode of growth could mean immortality; indeed, the longest lived individual organisms ever to have existed on earth have been certain species of trees. Plants and plant parts, however, do die, and death is often not the consequence of accident or environmental stress but of physiological decline—aging, or senescence.

Human Development

Developmental biology research has a long history of studying model organisms such as the fruit fly Drosophila melanogaster, focusing mainly on early embryogenesis. One major revolution in the field came in the early 1990s, when knockout mouse technology led to a focus on embryogenesis and organ development in mammals. Now, recent advances in human embryonic stem cell (ESC) and induced pluripotent stem cell (iPSC) technology have enabled the generation of human organoids (miniature versions of human tissues and organs grown in culture) of a variety of tissues, which can at least partially mimic human diseases and potentially be used in drug discovery platforms. Genome editing is further accelerating these research activities. In addition, single-cell sequencing is becoming a powerful tool for developmental biology, revealing the states of individual cells during embryogenesis and hence allowing investigation of the dynamics of tissue and organ development in greater detail. This technology is now being applied to human embryos, as well organoids generated in vitro. Human embryos are obtained and analyzed routinely at some institutions to study human development, not only for RNA sequencing but also for live imaging. These technological advances made for a very timely symposium, although the scope of this meeting was in fact much broader than our expectations. Rapid progress in next-generation sequencing has enabled comparative studies between humans and other mammals that are revealing the evolutionary processes that led to modern humans. Thus, we are now in the era of human developmental biology.

Development and Organogenesis

The process by which an organism develops from a single-celled zygote to a multi-cellular organism is complex and well regulated. The regulation occurs through signaling between cells and tissues and responses in the form of differential gene expression.

Early Embryonic Development

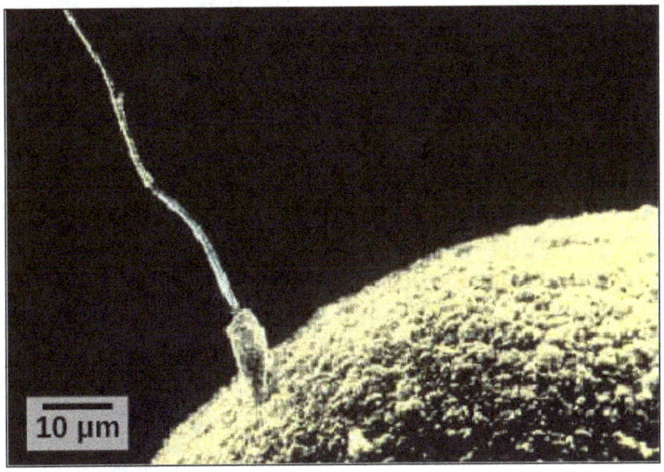

Fertilization is the process in which sperm and egg fuse to form a zygote.

Fertilization is the process in which gametes (an egg and sperm) fuse to form a zygote. To ensure that the offspring has only one complete diploid set of chromosomes, only one sperm must fuse with one egg. In mammals, a layer called the zona pellucida protects the egg. At the tip of the head of a sperm cell is a structure like a lysosome called the acrosome, which contains enzymes? When a sperm binds to the zona pellucida, a series of events, called the acrosomal reactions, take place. These reactions, involving enzymes from the acrosome, allow the sperm plasma membrane to fuse with the egg plasma membrane and permit the sperm nucleus to transfer into the ovum. The nuclear membranes of the egg and sperm break down and the two haploid nuclei fuse to form a diploid nucleus or genome.

To ensure that no more than one sperm fertilizes the egg, once the acrosomal reactions take place at one location of the egg membrane, the egg releases proteins in other locations to prevent other sperm from fusing with the egg.

The development of multi-cellular organisms begins from this single-celled zygote, which undergoes rapid cell division, called cleavage, to form a hollow ball of cells called a blastula.

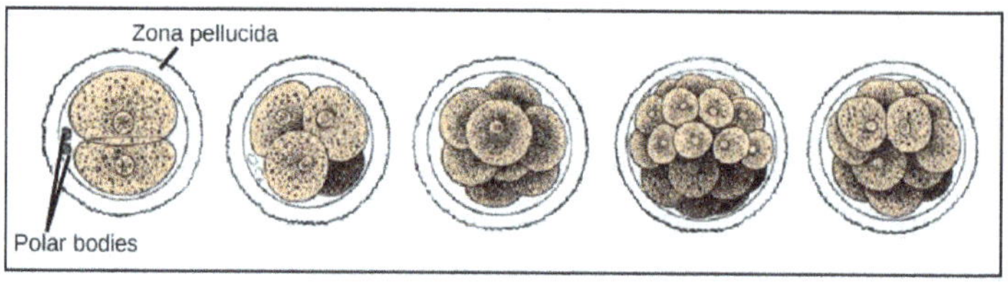

(a)

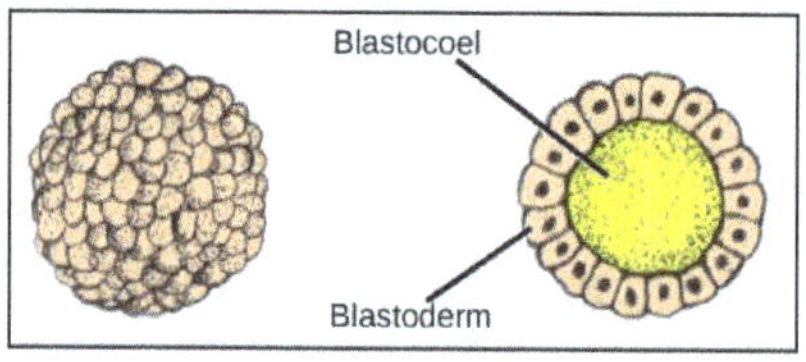

(b)

(a) During cleavage, the zygote rapidly divides into multiple cells. (b) The cells rearrange themselves to form a hollow ball called the blastula.

In mammals, the blastula forms the blastocyst in the next stage of development. Here the cells in the blastula arrange themselves in two layers: the inner cell mass and an outer layer called the trophoblast. The inner cell mass will go on to form the embryo. The trophoblast secretes enzymes that allow implantation of the blastocyst into the endometrium of the uterus. The trophoblast will contribute to the placenta and nourish the embryo.

The cells in the blastula then rearrange themselves spatially to form three layers of cells. This process is called gastrulation. During gastrulation, the blastula folds in on itself and cells migrate to form the three layers of cells in a structure, the gastrula, with a hollow space that will become the digestive tract. Each of the layers of cells is called a germ layer and will differentiate into different organ systems.

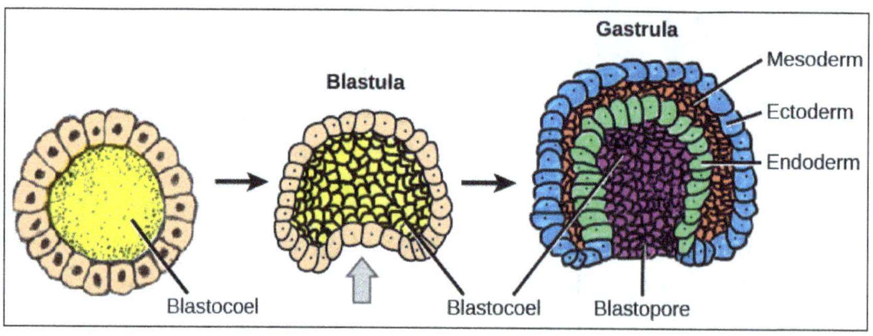

Gastrulation is the process wherein the cells in the blastula rearrange themselves to form the germ layers.

The three germ layers are the endoderm, the ectoderm, and the mesoderm. Cells in each germ layer differentiate into tissues and embryonic organs. The ectoderm gives rise to the nervous system and the epidermis, among other tissues. The mesoderm gives rise to the muscle cells and connective tissue in the body. The endoderm gives rise to the gut and many internal organs.

Organogenesis

Gastrulation leads to the formation of the three germ layers that give rise during further development to the different organs in the animal body. This process is called organogenesis.

Organs develop from the germ layers through the process of differentiation. During differentiation, the embryonic stem cells express specific sets of genes that will determine their ultimate cell type. For example, some cells in the ectoderm will express the genes specific to skin cells. As a result, these cells will take on the shape and characteristics of epidermal cells. The process of differentiation is regulated by location-specific chemical signals from the cell's embryonic environment that sets in play a cascade of events that regulates gene expression.

Genetic Variation and Inheritance

Genetic variation, the genetic difference between individuals, is what contributes to a species' adaptation to its environment. In humans, genetic variation begins with an egg, several million sperm, and fertilization. The egg and the sperm each contain 23 chromosomes, which make up our genes. A single gene may have multiple possible variations or alleles (a specific version of a gene), resulting in a variety of combinations of inherited traits.

Genetic inheritance of traits for humans is based upon Gregor Mendel's model of inheritance. For genes on an autosome (any chromosome other than a sex chromosome), the alleles and their associated traits are autosomal dominant or autosomal recessive. In this model, some genes are considered dominant because they will be expressed. Others, termed recessive, are only expressed in the absence of a dominant gene. Some characteristics which were once thought of as dominant-recessive, such as eye color, are now believed to be a result of the interaction between several genes. Dominant traits include curly hair, facial dimples, normal vision, and dark hair. Recessive characteristics include red hair, pattern baldness, and nearsightedness.

Sickle cell anemia is an autosomal recessive disease; Huntington disease is an autosomal dominant disease. Other traits are a result of partial dominance or co-dominance in which both genes are influential. For example, if a person inherits both recessive genes for cystic fibrosis, the disease will occur. But if a person has only one recessive gene for the disease, the person would be a carrier of the disease.

In this example, we will call the normal gene "N," and the gene for cystic fibrosis "c." The normal gene is dominant, which means that having the dominant allele either from one parent (Nc) or both parents (NN) will always result in the phenotype associated with the dominant allele. When someone has two copies of the same allele, they are said to be homozygous for that allele. When someone has a combination of alleles for a given gene, they are said to be heterozygous. For example, cystic fibrosis is a recessive disease which means that an individual will only have the disease if they are homozygous for that recessive allele (cc).

Imagine that a woman who is a carrier of the cystic fibrosis gene has a child with a man who also is a carrier of the same disease. What are the odds that their child would inherit the disease? Both the woman and the man are heterozygous for this gene (Nc). We can expect the offspring to have a 25% chance of having cystic fibrosis (cc), a 50% chance of being a carrier of the disease (Nc), and a 25% chance of receiving two normal copies of the gene (NN).

Where do harmful genes that contribute to diseases like cystic fibrosis come from? Gene mutations provide one source of harmful genes. A mutation is a sudden, permanent change in a gene. While many mutations can be harmful or lethal, once in a while a mutation benefits an individual by giving that person an advantage over those who do not have the mutation. Recall that the theory of evolution asserts that individuals best adapted to their particular environments are more likely to reproduce and pass on their genes to future generations. In order for this process to occur, there must be competition—more technically, there must be variability in genes (and resultant traits) that allow for variation in adaptability to the environment. If a population consisted of identical individuals, then any dramatic changes in the environment would affect everyone in the same way, and there would be no variation in selection. In contrast, diversity in genes and associated traits allows some individuals to perform slightly better than others when faced with environmental change. This creates a distinct advantage for individuals best suited for their environments in terms of successful reproduction and genetic transmission.

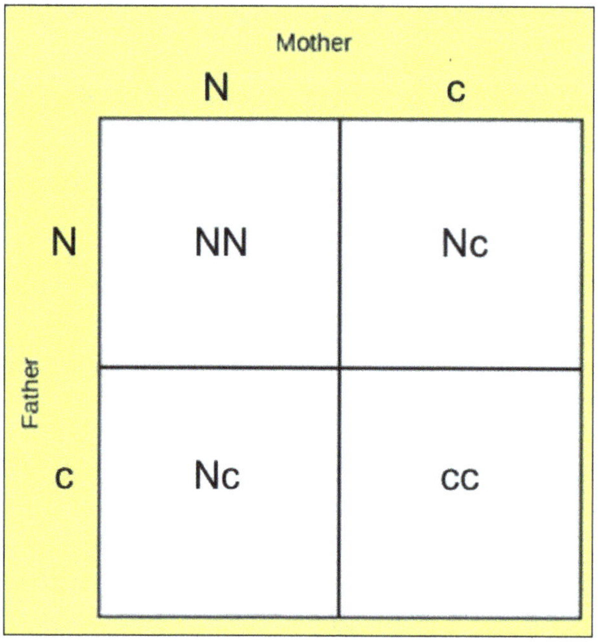

Figure shows a Punnett square is a tool used to predict how genes will interact in the production of offspring. The capital N represents the dominant allele, and the lowercase c represents the recessive allele. In the example of the cystic fibrosis, where N is the normal gene (dominant allele), wherever a pair contains the dominant allele, N, you can expect a phenotype that does not express the disease. You can expect a cystic fibrosis phenotype only when there are two copies of the c (recessive allele) which contains the gene mutation that causes the disease.

Animal Development

Animal development is the processes that lead eventually to the formation of a new animal starting from cells derived from one or more parent individuals. Development thus occurs following the process by which a new generation of organisms is produced by the parent generation.

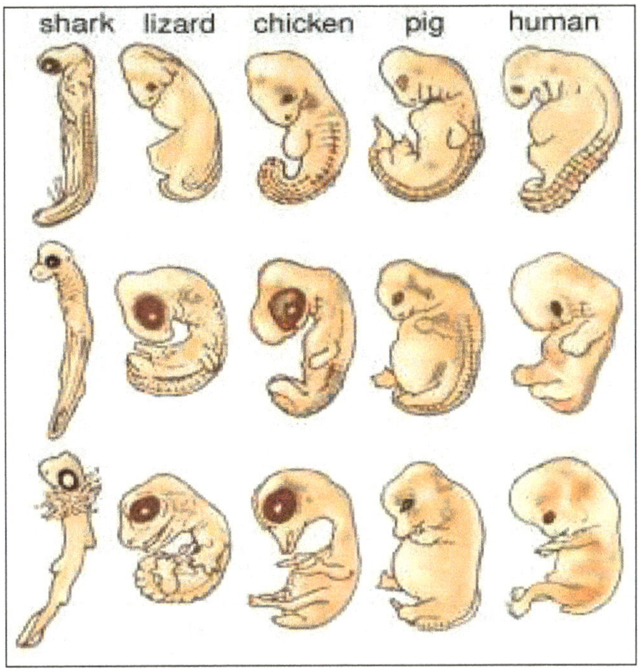

Embryos of different animals: The embryos of many animals appear similar to one another in the earliest stages of development and progress into their specialized forms in later stages.

Reproduction and Development

In multicellular animals (Metazoa), reproduction takes one of two essentially different forms: sexual and asexual. In asexual reproduction the new individual is derived from a blastema, a group of cells from the parent body, sometimes, as in Hydra and other coelenterates, in the form of a "bud" on the body surface. In sponges and bryozoans, the cell groups from which new individuals develop are formed internally and may be surrounded by protective shells; these bodies, which may serve as resistant forms capable of withstanding unfavourable environmental conditions, are released after the death of the parent. In certain animals the parent may split in half, as in some worms, in which an individual worm breaks into two fairly equal parts (except that the anterior half receives the mouth, "brain," and sense organs if they are present).

Obviously, in such a case it is impossible to say which of the two resulting individuals is the parent and which the offspring. Some brittle stars (starfish relatives) may reproduce by breaking across the middle of the body disk, with each of the halves subsequently growing its missing half and the corresponding arms.

A common feature of all forms of asexual reproduction is that the cells—always a substantial number of cells, never only one cell—taking part in the formation of the new individual are not essentially different from other body, or somatic, cells. The number of chromosomes (bodies carrying the hereditary material) in the cells participating in the formation of a blastema is the same as in the other somatic cells of the parent, constituting a normal, double, or diploid (2n), set.

In sexual reproduction, a new individual is produced not by somatic cells of the parent but by sex cells, or gametes, which differ essentially from somatic cells in having undergone meiosis, a process in which the number of chromosomes is reduced to one-half of the diploid (2n) number found in somatic cells; cells containing one set of chromosomes are said to be haploid (n). The resulting sex cells thus receive only half the number of chromosomes present in the somatic cell. Furthermore, the sex cells are generally capable of developing into a new individual only after two have united in a process called fertilization.

Each type of reproduction—asexual and sexual—has advantages for the species. Asexual reproduction is, at least in some cases, the faster process, leading most rapidly to the development of large numbers of individuals. Males and females are independently capable of producing offspring. The large size of the original mass of living matter and its high degree of organization—the new individual inherits parts of the body of the parent: a part of the alimentary canal, for instance—make subsequent development more simple, and the attainment of a stage capable of self-support easier. New individuals produced by asexual reproduction have the same genetic constitution (genotype) as their parent and constitute what is called a clone. Though asexual reproduction is advantageous in that, if the parent animal is well adapted to its environment and the latter is stable, then all offspring will benefit, it is disadvantageous in that the fixed genotype not only makes any change in offspring impossible, should the environment change, but also prevents the acquisition of new characteristics, as part of an evolutionary process. Sexual reproduction, on the other hand, provides possibilities for variation among offspring and thus assists evolution by allowing new pairs of genes to combine in offspring. Since all body cells are derived from the fertilized egg cell, a mutation, or change, occurring in the sex cells of the parents immediately provides a new genotype in each cell of the offspring. In the course of evolution, sexual reproduction has been selected for, and established in, all main lines of organisms; asexual reproduction is found only in special cases and restricted groups of organisms.

Preparatory Events

In the case of multicellular animals we find there are two kinds of sex cells: the female sex cell (ovum, or egg), derived from an oocyte (immature egg), and the male sex cell (spermatozoon or sperm), derived from a spermatocyte. Eggs are produced in ovaries; sperm, in testes. Both the egg and the sperm contribute to the development of the new individual; each providing one set of genes, thereby restoring the diploid number of chromosomes in the fertilized egg. The sperm possesses a whiplike tail (flagellum) that enables it to swim to the egg to fertilize it. In most cases the egg, a stationary, spherical cell, provides the potential offspring with a store of food materials, or yolk, for its early development. The term yolk does not refer to any particular substance but in fact includes proteins, phosphoproteins, lipids, cholesterol, and fats, all of which substances occur in various proportions in the eggs of different animals. In addition to yolk, eggs accumulate other components and acquire the structure necessary for the development of the new individual. In particular the egg acquires polarity—that is, the two ends, or poles, of the egg become distinctive from each other. At one pole, known as the animal pole, the cytoplasm appears to be more active and contains the nucleus (meiotic divisions occur in this region); at the other, called the vegetal pole, the cytoplasm is less active and contains most of the yolk. The general organization of the future animal is closely related to the polarity of the egg.

When the amount of food reserve is comparatively small, as it is in many marine invertebrates and mammals (in the latter the embryo is nourished by materials in the mother's blood), the egg may be barely visible to the unaided eye. The egg of the sea urchin is about 75 microns (0.003 inch) in diameter; that of a human being is slightly more than 0.1 millimetre. Eggs are classified according to the amount of yolk present. An egg with a small quantity of evenly distributed yolk is called an oligolecithal egg. One with more yolk that is unevenly distributed (i.e., concentrated towards the vegetal pole) is telolecithal; and one with still greater amounts of yolk in granules or in a compact mass is megalecithal.

The egg is surrounded by protective membranes, which may be soft and jellylike or hard and calcified, like shells. Egg membranes are produced while the egg is either in the ovary or being carried away from the ovary in a tube called an oviduct. The eggs of many animals have both kinds of membranes. In insects, a hard shell (chorion) forms around the eggs in the ovaries. In frogs, a very thin vitelline membrane forms around the eggs in the ovary; subsequently a layer of jelly is deposited around the eggs while they pass through the oviducts. In birds, a very thin vitelline membrane is produced around the egg in the ovary; then several layers of secondary membranes are formed in the oviduct before the egg is laid. The outermost of these secondary membranes is the calcareous shell. In mammals the egg is surrounded by the so-called pellucid zone, which is equivalent to the vitelline membrane of other animals; follicle cells form an area called the corona radiata around this zone.

After fertilization the egg, now called a zygote, is endowed with genes from two parents and has begun actual development. Activation of the egg may be brought about by an agent other than sperm in certain animals, but such cases of parthenogenesis are exceptional.

After fertilization, the zygote undergoes a series of transformations that bring it closer to the essential organization of the parents. These transformations, initiated at a physiological, perhaps even at a molecular, level, eventually result in the appearance of certain structures. The whole process is called morphogenesis. The process of development is more easily understood if, at every step, the changes necessary to bring the system nearer the goal are considered. Depending on the achievements necessary at any step, development can be subdivided into a number of discrete phases, the first of which, cleavage, immediately follows fertilization.

Embryo Formation

Cleavage

Since the goal of development is the production of a multicellular organism, many cells must be produced from the single-celled zygote. This task is accomplished by cleavage, a series of consecutive cell divisions. Cells produced during cleavage are called blastomeres. The divisions are mitotic—i.e., each chromosome in the nucleus splits into two daughter chromosomes, so that the two daughter blastomeres retain the diploid number of chromosomes. During cleavage, almost no growth occurs between consecutive divisions, and the total volume of living matter does not change substantially; as a consequence, the size of the cells is reduced by almost half at each division. At the beginning of cleavage, cell divisions tend to occur at the same time in all blastomeres, and the number of cells is doubled at each division. As cleavage progresses, the cells no longer divide at the same time.

Cleavage in most animals follows an orderly pattern, with the first division being in the plane of the main axis of the egg. This cleavage plane is arbitrarily called vertical, on the assumption that the main axis of the egg is vertical. The second cleavage plane is again vertical but at right angles to the first, giving rise to four equal cells arranged around the main axis of the egg. The third cleavage plane is at right angles to both the first and second cleavage planes and is horizontal, or equatorial. Subsequent divisions may alternate between vertical and horizontal cleavage planes, but later cleavage divisions become randomly oriented. This pattern is typical of many animal groups; however, more complicated patterns of cleavage are found in such animals as annelids, mollusks, and nematodes.

As the amount of yolk in the egg increases, it influences cleavage by hindering the cytoplasmic movements involved in mitosis. If there is only little yolk (oligolecithal eggs), the yolk granules follow the movements of the cytoplasm and are distributed in the resulting blastomeres. But if the amount of yolk is larger (megalecithal eggs), cleavages occur nearer the animal pole, where there is less yolk; as a result, the blastomeres nearer the animal pole are smaller than those nearer the vegetal pole. The presence of yolk masses may retard the onset of cleavage in a part of the egg or even suppress it altogether; in this case cleavage is partial, or meroblastic. Only a part of the egg material then is subdivided into cells, the rest remaining as a mass that serves as nourishment for the developing embryo.

Cleavage is complete, or holoblastic, in many invertebrates including coelenterates, annelids, echinoderms, tunicates, and cephalochordates. The blastomeres may be either about equal or only slightly different in size. Cleavage in amphibians is holoblastic, but the size of the blastomeres is very uneven. Blastomeres are smallest at the animal pole and largest (and yolky) at the vegetal pole. Somewhat similar conditions prevail in many mollusks. In most fishes, birds, reptiles, and egg-laying mammals (monotremes), cleavage is discoidal—i.e., restricted to a disk of cytoplasm at the animal pole of the egg, most of the yolky egg material remaining uncleaved. Cleavage in insects and many other arthropods is superficial—i.e., the entire surface layer of egg cytoplasm subdivides into cells, and the egg contains a central mass of uncleaved yolk. The conditions of cleavage in placental mammals, including man, are peculiar.

During cleavage, development involves only an increase of cell numbers; the shape of the embryo does not change, and chemical transformations within the embryo are restricted to those necessary for cell division. Chemical and structural transformations are concerned with accumulating chromosomal material in the nuclei of the blastomeres. Before each division the chromosomes carrying the genes double in number; this means that the chromosomal material, deoxyribonucleic acid (DNA), has to be synthesized. This synthesis proceeds possibly at the expense of cytoplasmic ribonucleic acid (RNA) but certainly also from simpler organic compounds. A certain amount of protein synthesis is also necessary for cleavage to proceed: if developing eggs are treated with puromycin, a substance which is known to suppress protein synthesis, cleavage stops immediately. The proteins concerned have not yet been identified. No proteins are synthesized, however, that would foreshadow the future differentiation of parts of the embryo. It is believed that the genes in the chromosomes remain largely inactive during cleavage. The rhythm (speed) of cleavage is wholly dependent on the cytoplasm of the egg.

Although the shape and volume of the embryo do not change during cleavage, one important change in gross organization does take place. As the blastomeres are produced, they move outward, leaving a centrally located fluid-filled cavity. In cases of holoblastic cleavage, the blastomeres become arranged in a layer from one to several cells thick surrounding the cavity. The embryo at this stage may be likened to a hollow ball and is known as a blastula. The outer layer of cells is called the blastoderm, and the fluid-filled cavity the blastocoel. In discoidal cleavage the cells, which do not surround the whole embryo, lie only on the animal pole; nevertheless, a blastocoel may be formed by a crevice appearing between the blastomeres and the mass of yolk. The blastomeres then may be arranged as a saucer-shaped blastodisk covering the blastocoel.

The formation of the blastula signifies the end of the period of cleavage. The next stage of development is concerned not with an increase in cell number, though cell divisions continue at a slower pace, but with rearrangement of the available cell masses to conform to the gross features of the future animal.

Gastrulation

The embryo in the blastula stage must go through profound transformations before it can approach adult organization. An adult multicellular animal typically possesses a concentric arrangement of tissues of the body; this feature is common to all animal groups above the level of the sponges. Adult tissues are derived from three embryonic cell layers called germinal layers: the outer layer is the ectoderm, the middle layer is the mesoderm, and the innermost layer is the endoderm (entoderm). The ectoderm gives rise to the skin covering, to the nervous system, and to the sense organs. The mesoderm produces the muscles, excretory organs, circulatory organs, sex organs (gonads), and internal skeleton. The endoderm lines the alimentary canal and gives rise to the organs associated with digestion and, in chordates, with breathing.

The blastula, which consists of only one cell layer, undergoes a dramatic reshuffling of blastomeres preparatory to the development of the various organ systems of the animal's body. This is achieved by the process of gastrulation, which is essentially a shifting or moving of the cell material of the embryo so that the three germinal layers are aligned in their correct positions.

The rearrangement of the blastula to form the germinal layers is seen clearly in certain marine animals with oligolecithal eggs. The hollow blastula consists of a simple epithelial layer (the blastoderm), the transformation of which can be likened to the pushing in of one side of a rubber ball. As a result of such inpushing (or invagination), the spherical embryo is converted into a double-walled cup, the opening of which represents the position of the former vegetal pole. The involuted part of the blastoderm, lining the inside of the double-walled cup, gives rise to the endoderm and mesoderm, and the blastomeres remaining on the exterior become the ectoderm. As a consequence of the infolding at the vegetal pole, the blastocoel is reduced or obliterated, and a new cavity is created, the primitive gut cavity, or archenteron, which eventually gives rise to the hollow core (lumen) of the alimentary canal. At this stage the embryo has a primitive gut with an opening to the exterior and is known as a gastrula. The opening of the gastrula is the blastopore, or primitive mouth; both terms are somewhat misleading. It would seem that the term blastopore should be applied more appropriately to an opening in a blastula, in which, of course, no opening exists. As to the term primitive mouth, it must be pointed out that the blastopore does not always give rise to the adult mouth. In certain animal groups it becomes the anus, and a mouth forms as a completely new opening.

In some coelenterates, cells at the vegetal pole do not form an invaginating pocket, but individual cells slide inward, losing connection with other cells of the blastoderm. Eventually these cells fill the blastocoel and form a compact mass of endoderm. The cavity (archenteron) within this mass and the opening (blastopore) to the exterior are then produced secondarily by the separation of these cells.

Amphioxus, Echinoderms and Amphibians

Gastrulation does not always proceed exactly as described above. In the course of evolution, certain animal groups have modified this critical stage of embryonic development, and these modifications have undoubtedly contributed to the successful continuation of species. In the primitive fishlike chordate amphioxus, for example, the invaginating blastoderm eventually comes into close contact with the inner surface of the ectoderm, thus practically squeezing the blastocoel out of existence or at least reducing it to a narrow crevice between the ectoderm and the endomesoderm. In echinoderms, on the other hand, a smaller portion of the blastoderm invaginates, and the blastocoel remains as a spacious internal cavity between the ectoderm and the endomesoderm. It persists as the primary body cavity and is the only body cavity (apart from the cavity of the alimentary canal) in such invertebrates as nematodes and rotifers.

In the double-walled-cup stage, the two internal germinal layers—endoderm and mesoderm—may not yet be distinct. Their separation may occur later, in the second phase of gastrulation, by one of two methods. One is the development of outpocketings from the wall of the archenteron. In starfishes and other echinoderms, the deep part of the endomesodermal invagination forms two thin-walled sacs, one on each side of the gastrula. These are the rudiments of the mesoderm; the remaining part of the archenteron becomes the endoderm and produces the lining of the gut. The cavities within the mesodermal sacs expand to become the coelom, the secondary body cavity of the animal. A somewhat similar process of mesoderm and coelom development occurs in amphioxus among the chordates, except that a series of mesodermal sacs forms on either side of the embryo, foreshadowing the segmented (metameric) structure common to chordates. Only the most anterior pairs of the mesodermal sacs actually contain a cavity at the time of their formation; the more posterior ones are solid masses of cells separating from the archenteric wall and from one another and developing coelomic cavities later.

A second method of mesoderm formation is by the splitting off of mesodermal cells from the original common mass of endomesoderm. This may take the form of single cells detaching themselves from the archenteron or of whole sheets of cells splitting off from the endoderm. An example of the latter type is seen in the gastrulation of amphibians. The development of specific regions of the early amphibian embryo—by the use of natural pigmentation or artificially introduced dyes—can be followed and their location in the adult recorded in diagrams called fate maps. The fate map of a frog blastula just prior to gastrulation demonstrates that the materials for the various organs of the embryo are not yet in the position corresponding to that in which the organs will lie in a fully developed animal. The endodermal material for the foregut, for example, lies not far from the vegetal pole; the ectodermal component of the mouth region (stomodeum) is situated close to the animal pole. Extensive rearrangement of the embryo is necessary to bring all the parts into their correct relationships.

Because of the large amount of yolk and resulting uneven cleavage, gastrulation in amphibians cannot proceed by a simple infolding of the vegetal hemisphere. A certain amount of invagination does take place, assisted by an active spreading of the animal hemisphere of the embryo; as a result, the ectoderm covers the endodermal and mesodermal areas. The spreading is sometimes described as an "overgrowth"—an inappropriate term, since no growth or increase of mass is involved. The future ectoderm simply thins out, expands, and covers a greater surface of the embryo in a movement known as epiboly.

Gastrulation in amphibians, in lungfishes, and in the cyclostomes (hagfishes and lampreys) begins with the formation of a pit on what will become the back (dorsal) side of the embryo. The pit represents the active shifting inward of the cells of the blastoderm. As these cells undergo a change in shape, there occurs also a contraction at the external surface, with adjacent cells being drawn toward the centre of the contraction even before an actual depression is formed. The cells most concerned in this process will become part of the future foregut. Further movement of the cells inward results in the formation of a distinct pit, which rapidly develops into a pocket-like archenteron with its opening, the blastopore. Once the archenteron is formed, more and more of the exterior cells roll over the edge of the blastopore and disappear into the interior. In the course of gastrulation the shape of the blastopore changes from a simple pit to a transverse slit and finally into a groove encircling the yolky material at the vegetal pole. As a result of epiboly of the animal hemisphere, the upper edge of the groove is gradually pushed down until the yolky cells of the vegetal pole are covered completely. The edges of the blastopore then converge toward the vegetal pole, the slit between them being eventually reduced to a narrow canal, which lies at the posterior end of the embryo and, in some species, becomes the anal opening. (In other cases the canal closes, and a new anal opening breaks through nearby, slightly more ventrally.)

The cavity of the archenteron increases as more material from the outside is transferred inward, and the blastocoel becomes almost completely obliterated. Both mesoderm and endoderm are shifted into the interior, and only the ectoderm remains on the embryo surface. The mesoderm splits from the endoderm: the endoderm lines the archenteric cavity (and eventually becomes the lining of the alimentary canal), as the mesoderm surrounds the endoderm to form the chordamesodermal mantle. By the time the blastopore closes, the three germ layers are in their correct spatial relationship to each other.

Reptiles, Birds and Mammals

Although amphibian gastrulation is considerably modified in comparison with that in animals with oligolecithal eggs (e.g., amphioxus and starfishes), an archenteron forms by a process of invagination. Such is not the case, however, in the higher vertebrates that possess eggs with enormous amounts of yolk, as do the reptiles, birds, and egg-laying mammals. Cleavage in these animals is partial (meroblastic), and, at its conclusion, the embryo consists of a disk-shaped group of cells lying on top of a mass of yolk. This cell group often splits into an upper layer, the epiblast, and a lower layer, the hypoblast. These layers do not represent ectoderm and endoderm, respectively, since almost all the cells that form the embryo are contained in the epiblast. Future mesodermal and endodermal cells sink down into the interior, leaving only the ectodermal material at the surface. In reptiles, egg-laying mammals, and some birds, a pocket-like depression occurs in the epiblast but encompasses only chordamesoderm or even only the notochord. Individual cells of the remainder of the mesoderm and endoderm migrate into the interior and there arrange themselves into a sheet of chordamesoderm and of endoderm, the latter of which mingles with cells of the hypoblast if such a layer is present. The migration of the cells destined to form mesoderm and endoderm does not take place over the whole surface of the disk-shaped embryo but is restricted to a specific area along the midline. This area is more or less oval in reptiles and lower mammals; distinctly elongated in higher mammals and birds, it is called the primitive streak, a thickened and slightly depressed part of the epiblast that is thickest at the anterior end, called the Hensen's node.

In animals having discoidal cleavage, the three germinal layers at the end of gastrulation are stacked flat; ectoderm on top, mesoderm in the middle, and endoderm at the bottom. The embryo is produced from the flattened layers by a process of folding to form a system of concentric tubes. The edges of the germ layers, which are not involved in the folding process, remain attached to the yolk and become the extra-embryonic parts; they are not directly involved in supplying cells for the embryo but break down yolk and transport it to the developing embryo.

Higher mammals—apart from the egg-laying mammals—do not have yolk in their eggs but, having passed through an evolutionary stage of animals with yolky eggs, retain, particularly in gastrulation, features common to reptiles (and birds, which also had reptilian ancestors). As a result, at the end of cleavage the formative cells of the embryo—the cells that will actually build the body of the animal—are arranged in the form of a disk over a cavity that takes the place of the yolk of the reptilian ancestors of mammals. Within the disk of cells a primitive streak develops, and the three germinal layers are formed much as in many reptiles and birds.

Gastrulation and the formation of the three germinal layers is the beginning of the subdivision of the mass of embryonic cells produced by cleavage. The cells then begin to change and diversify under the direction of the genes. The genes brought in by the sperm exert control for the first time; during cleavage all processes seem to be under control of the maternal genes. In cases of hybridization, in which individuals from different species produce offspring, the influence of the sperm is first apparent at gastrulation: paternal characteristics may appear at this stage; or the embryo may stop developing and die if the paternal genes are incompatible with the egg (as is the case in hybridization between species distantly related).

The diversification of cells in the embryo progresses rapidly during and after gastrulation. The visible effect is that the germinal layers become further subdivided into aggregations of cells that assume the rudimentary form of various organs and organ systems of the embryo. Thus the period of gastrulation is followed by the period of organ formation, or organogenesis.

Embryonic Adaptations

Throughout its development the embryo requires a steady supply of nourishment and oxygen and a means for disposal of wastes. These needs are met in various ways, depending in particular on (1) whether the eggs develop externally (oviparity), are retained in the maternal body until ready to hatch (ovoviviparity), or are carried in the maternal body to a later stage (viviparity); and (2) the length of embryonic development.

Adaptations in Animals other than Mammals

Eggs of many marine invertebrates are discharged directly into water, and the period of development before the larva emerges is relatively brief. Oxygen diffuses easily into the small eggs, and nourishment is provided by a moderate amount of yolk. During cleavage the yolk is distributed to all the blastomeres. Much of the nourishment in the egg is stored as animal starch, or glycogen, which is almost completely used by the time the larva emerges from the egg. A small amount of water and inorganic salts are taken in by the embryo from surrounding seawater. Eggs developing in freshwater carry their own supply of necessary amounts of certain salts that are not present in sufficient quantities in the environment. Products of metabolism—especially carbon dioxide and nitrogenous wastes in the form of ammonia—diffuse out from small embryos developing in water.

The eggs of terrestrial animals must overcome the hazard of drying. In certain species this danger is avoided because the animal returns to water to breed, such as frogs and salamanders. Some groups of insects (e.g., dragonflies, mayflies, and mosquitoes) also lay eggs in water, and the larvae are aquatic. Eggs of other animals (e.g., snails, earthworms) are laid in moist earth and thus are protected from drying up. In terms of evolution, however, a decisive solution to the problem of development on land was arrived at by most insects and by reptiles and birds, which developed eggs with a shell impermeable to water or, at least, resistant to rapid evaporation. The shells of bird and insect eggs, while restricting evaporation of water, allow oxygen to diffuse into the egg and carbon dioxide to diffuse out. Apart from gas exchange, the eggs constitute closed systems, which give nothing to the outside and require nothing from it. Such eggs are called cleidoic. Because the products of nitrogen metabolism in cleidoic eggs cannot pass through the eggshell, animals (birds and insects) have had to evolve a method of storing wastes in the form of uric acid, which, since it is insoluble, is nontoxic to the embryo.

After a short period of development in the egg, the emerging young animal has to fend for itself, unless there is some form of parental care. Exposure to the external environment at a tender age results frequently in loss of life, a hazard met by many animals through an increase in the supply of nourishment within the egg, thus allowing the young to attain a greater size and development. This tendency to produce large yolky eggs has been achieved independently in different evolutionary lines: in octopuses and squids among the mollusks, in sharks among the fishes, and in reptiles and birds among the terrestrial vertebrates.

As has been indicated, cleavage is incomplete in eggs with large amounts of yolk. Although some yolk platelets may be enclosed in the formative cells of the embryo, the bulk of the yolk remains an uncleaved mass, overgrown and surrounded by the cellular part of the embryo. In such cases a membranous bag, or yolk sac, is formed and remains connected to the embryo by a narrow stalk (the evolutionary precursor of the umbilical cord of mammals). The cellular layers surrounding the yolk sac and forming its walls may consist of all three germinal layers (in reptiles and birds), so that the yolk virtually comes to lie inside an extension of the gut of the embryo; or (in bony fishes) the yolk sac may be enclosed in layers of ectoderm and mesoderm. In either case a network of blood vessels develops in the walls of the yolk sac and transports the yolk products to the embryo. As the yolk is broken down and utilized, the yolk sac shrinks and is eventually drawn into the body of the embryo. In addition to the yolk sac, extra-embryonic parts are also encountered in the form of embryonic membranes, which are found in higher vertebrates and in insects. Vertebrates have three embryonic membranes: the amnion, the chorion, and the allantois.

In reptiles, birds, and mammals, folds develop on the surface of the yolk sac just outside and around the body of the embryo proper. These folds, consisting of extra-embryonic ectoderm and extra-embryonic mesoderm, rise up and fuse dorsally, enclosing the embryo in a double-lined, fluid-filled chamber known as the amniotic cavity. The inner lining of the fold becomes the amnion, and the outer becomes the chorion, which ultimately surrounds the entire embryo. The amniotic fluid protects the embryo from drying, prevents the adhesion of the embryo to the inner surface of the shell, and provides the embryo with efficient shock absorption against possible damaging jolts. (The aminion and chorion develop in the same way in insect embryos.) The third membrane, or allantois, is originally nothing more than the urinary bladder of the embryo. It is a saclike growth of the floor of the gut, into which nitrogenous wastes of the embryo are voided. It enlarges greatly during the course of development, eventually expanding between the amnion and chorion and also between the chorion and the yolk sac, to become the third embryonic membrane. In addition to providing storage space for the nitrogenous wastes of the embryo, the allantois takes up oxygen, which penetrates into the egg from the exterior, and delivers it, by way of a network of blood vessels, to the embryo.

Adaptations in Mammals

At some early stage during the evolution of viviparous mammals, eggs came to be retained in the oviducts of the mother. The embryo then was provided with nourishment from fluids in the oviduct; the yolk, which became redundant, gradually ceased to be provided, and the eggs became oligolecithal. The eggshell, present in reptiles, was no longer needed and eventually disappeared, as did the white of the egg. The chorion, however, remained as the most external coat of the developing embryo through which nourishment reaches the embryo. It acquired the ability to adhere closely to the walls of the uterus (which was what that part of the oviduct holding the embryo had become) and became the so-called trophoblast. The blood-vessel network of the underlying allantois conveys nutrients that diffuse through the trophoblast to the body of the embryo proper. These modifications gave rise to a new organ, the placenta, formed from tissues of both the mother and the embryo: the uterine wall with its blood vessels provided by the mother; the trophoblast and allantois—and in some mammals also the yolk sac—with their blood vessels provided by the embryo.

The overall development of placental mammals as a result of these changes is profoundly different from that of their ancestors, the reptiles, and proceeds in the following way: the tiny yolkless egg is fertilized in the upper portion of the oviduct by sperm received from the male in the process of coupling (coitus); cleavage starts as the egg is propelled slowly down the oviduct by action of cilia in the oviduct lining. At the end of cleavage a solid ball of cells called a morula is produced. The surface cells of the morula become the trophoblast and the inner cell mass gives rise to the embryo (the formative cells) and also its yolk sac, amnion, and allantois. A cavity appears within the morula, converting it into a hollow embryo, called the blastocyst. This cavity resembles the blastocoel but, in fact, is analogous to the yolk sac of meroblastic eggs, except that there is no yolk and the cavity is filled with fluid. At the blastocyst stage, the embryo enters the uterus and attaches itself to the uterine wall. This attachment, or implantation, a crucial step in the development of a mammal, is attained through the action of the trophoblast, which forms extensions, known as villi that penetrate the uterine wall. In higher placental mammals, the lining of the uterine wall and, in varying degrees, the underlying tissues as well are partially destroyed, resulting in a closer relationship between the blood supplies of the mother and the embryo. Indeed, in man and in some rodents, the blastocyst sinks completely into the uterine wall and becomes surrounded by uterine tissue.

While implantation takes place, the formative cells arrange themselves in the form of a disk under the trophoblast. In the disk, the germinal layers develop much as in birds, with the formation of a primitive streak and migration of the chordamesoderm into a deeper layer. A layer of endoderm is formed adjoining the cavity of the blastocyst, and an amniotic cavity develops, enclosing the embryo; in lower placental mammals, the allantois also develops. The embryo proper, lying in the amniotic cavity, is connected to the extra-embryonic parts by the umbilical cord. The umbilical cord lengthens greatly during later development. In higher mammals, the cavity of the allantois is reduced, but the allantoic blood vessels become well developed and extend through the umbilical cord, connecting the embryo to the placenta. The blood that circulates in the placenta brings oxygen and nutrients from the maternal blood to the embryo and carries away carbon dioxide and other waste products from the embryo to the maternal blood for disposal by the maternal body.

Although tissues of maternal and embryonic origin are closely apposed in the placenta, there is little actual mingling of the tissues. Despite an occasional penetration of an embryo cell into the mother and vice versa, there is a placental barrier between the two tissues. The blood circulation of the mother is at all times completely separated from that of the embryo and its extra-embryonic parts. The placental barrier, however, does allow molecules of various substances to pass through; such differential permeability is indeed necessary if the embryo is to obtain nourishment. The permeability of the placental barrier differs in different animals; thus antibodies, which are protein molecules, may penetrate the placental barrier in man but not in cattle.

The maintenance of the fetus—as the more advanced embryo of a mammal is called—in the uterus is under hormonal control. In the initial stages of pregnancy, the continued existence of the embryo in the uterus depends on the hormone progesterone, which is secreted by the corpora lutea, "yellow bodies," that develop in the ovary after an egg has been released.

At birth the fetal parts of the placenta separate from the maternal parts. Contraction of the uterine wall first releases the fetus from the uterus; the fetal parts of the placenta (the afterbirth) follow. In certain cases of intimate connection between fetal and maternal tissues, the maternal tissues are torn, and birth is accompanied by profuse bleeding.

Organ Formation

Primary Organ Rudiments

Immediately after gastrulation—and sometimes even while gastrulation is underway—the germinal layers begin subdividing into regions that will give rise to various parts of the body. Subdivision proceeds in stages: initially a mass of cells is set aside for an organ system (for the alimentary canal, for instance) and subsequently further subdivided into the rudiments of various parts of the organ system, such as the liver, stomach, and intestines. The initially formed larger units are referred to as primary organ rudiments; those they later give rise to, as secondary organ rudiments.

Differentiation of the Germinal Layers

The type of organ rudiment produced depends on the organization of the body in any particular group in the animal kingdom. In the vertebrates the earliest subdivision within a germinal layer is the segregation within the chordamesodermal mantle of the rudiment of the notochord from the rest of the mesoderm. During gastrulation the material of the notochord comes to lay middorsally in the roof of the archenteron. It separates by longitudinal crevices from the chordamesodermal mantle lying to the left and right. The material of the notochord then rounds off and becomes a rod-shaped strand of cells immediately under the dorsal ectoderm, stretching from the blastopore toward the anterior end of the embryo, to the midbrain level. In front of the tip of the notochord, there remains a thin sheet of prechordal mesoderm.

The mesodermal layer adjoining the notochord becomes thickened and, by transverse crevices, subdivided into sections called somites. The somites, which later give rise to the segmented body muscles and the vertebral column, are the basis of the segmented organization typical of vertebrates (seen especially in the lower fishlike forms but also in the embryos of higher vertebrates). The lateral and ventral mesoderm, which remains unsegmented, is called the lateral plate. The somites remain connected to the lateral plate by stalks of somites that play a particular role in the development of the excretory (nephric) system in vertebrates; for this reason they are called nephrotomes. Rather early the mesodermal mantle splits into two layers, the outer parietal (somatic) layer and the inner visceral (splanchnic) layer, separated by a narrow cavity that will expand later to form the coelomic, or secondary, body cavity. The coelomic cavity extends initially through the nephrotomes into the somites; in the somites it is eventually obliterated. Endoderm completely surrounds the lumen of the archenteron (when present) and produces the cavity of the alimentary canal. If no archenteric cavity is formed during gastrulation, the cavity of the alimentary canal is formed by the separation of cells in the middle of the mass of endoderm (as in bony fishes) or by folding of the sheet of endoderm. The endodermal gut sooner or later acquires an extended anterior part called the foregut and a narrower and more elongated posterior part, the hindgut. Characteristic of chordates is the development of the nervous system from a part of ectoderm lying originally on the dorsal side of the embryo, above the notochord and the somites. This part of the ectodermal layer thickens and becomes the neural plate, whose edges rise as neural folds that converge toward the midline, fuse together, and form the neural tube. In vertebrates the neural tube lies immediately above the notochord and extends beyond its anterior tip. The neural tube is the rudiment of the brain and spinal cord; its lumen gives rise to the cavities, or ventricles, of the brain and to the central canal of the spinal cord. The remainder of the ectoderm closes over the neural tube and becomes, in the main, the covering layer (epithelium) of the animal's skin (epidermis). As the neural tube detaches itself from the overlying ectoderm, groups of cells pinch off and form the neural crest, which plays an important role in the development of, among other things, the segmental nerves of the brain and spinal cord.

In developing the primary organ rudiments mentioned above, the embryo acquires a definite organization clearly recognizable as that of a chordate animal. Similar processes, which occur in the development of other animals, establish the basic organization of an annelid, a mollusk, or an arthropod.

Embryonic Induction

The organization of the embryo as a whole appears to be determined to a large extent during gastrulation, by which process different regions of the blastoderm are displaced and brought into new spatial relationships to each other. Groups of cells that were distant from each other in the blastula come into close contact, which increases possibilities for interaction between materials of different origin. In the development of vertebrates in particular, the sliding of cells (presumptive mesoderm) into the interior and their placement on the dorsal side of the archenteron (in the archenteric "roof"), in immediate contact with the overlying ectoderm, is of major importance in development and subsequent differentiation. Experiments have shown that, at the start of gastrulation, ectoderm is incapable of progressive development of any kind; that only after invagination, with chordamesoderm lying directly underneath it, does ectoderm acquire the ability for progressive development. The dorsal mesoderm, which later differentiates into notochord, prechordal mesoderm, and somites, causes the overlying ectoderm to differentiate as neural plate. Lateral mesoderm causes overlying ectoderm to differentiate as skin. The influence exercised by parts of the embryo, which causes groups of cells to proceed along a particular path of development, is called embryonic induction. Though induction requires that the interacting parts come into close proximity, actual contact is not necessary. The inducing influence—whatever it might be—is a diffusible substance emitted by the activating cells (the inductor). The inducing substance of the mesoderm is a large molecule, probably a protein or a nucleoprotein, which presumably penetrates reacting cells, though direct and unequivocal proof of such penetration is still unavailable. Inducing substances are active on vertebrates belonging to many different classes; e.g., inductions of primary organs have been obtained by transplanting mammalian tissues into frog embryos or by transplanting tissues of a chick embryo into the embryo of a rabbit.

Induction is responsible not only for the subdivision of ectoderm into neural plate and epidermis but also for the development of a large number of organ rudiments in vertebrates. The notochord is a source of induction for the development of the adjoining somites and nephrotomes; the latter appear jointly to induce development of limb rudiments from the lateral plate mesoderm. Further examples are mentioned below in connection with development of the various organs.

Since the results of induction are different for different organ rudiments, it must be presumed that there exist inducing substances with specific action, at least to a certain extent; thus, the lateral mesoderm induces differentiation of the skin but not neural plate from the very same kind of ectoderm. The number of inducing substances need not, however, be the same as the number of different kinds of tissues and organs, since certain differentiations could possibly be induced by a combination of two or more inducing substances, or the same inducing substance might have different effects on different tissues. It has been suggested that the regional organization of the entire vertebrate body could be controlled by the graded distribution of only two inducing substances—provisionally named the neuralizing substance and the mesodermalizing substance—along the length of the embryo. The neuralizing substance, concentrated at the anterior end, gradually decreases toward the posterior end; the mesodermalizing substance, on the other hand, is concentrated at the posterior end and decreases toward the anterior end. The differentiation of induced structures depends on the relative amounts of the two inducing substances at any given point in the embryo. Acting alone, the neuralizing substance induces only nervous tissue, which takes the form of the forebrain, and the mesodermalizing substance induces only mesodermal structures (e.g., somites, notochord).

In the amphibian embryo, induction appears to have its primary source in the dorsal lip of the blastopore, which eventually gives rise to the notochord and adjoining somites. Induction by the notochord and somites is responsible for the development of the neural plate in the ectoderm, of lateral and ventral parts of the mesodermal mantle, and of the lumen of the alimentary canal in the endoderm. The dorsal lip of the blastopore for this reason has been called the primary organizer. In higher vertebrates, in which gastrulation occurs through the medium of a primitive streak, the anterior end of the streak and the Hensen's node have properties similar to those of a primary organizer. Organization centres have been found, or suspected, in embryos of animals belonging to a few other groups, in particular the insects and sea urchins, but the interpretation of the experimental results in these animals is less satisfactory than in the case of vertebrates.

The concept of an organization centre suggests that a part of the embryo differs from the rest of the embryonic tissues in being more active. The more active parts of the embryo (and also of animals in later stages of development) are particularly sensitive to certain noxious influences in their environment. If an embryo is deprived of oxygen or subjected to weak concentrations of poisons, the first parts to suffer are the most morphogenetically active ones. In vertebrate embryos the anterior end of the head is most sensitive. Early sea-urchin embryos have two centres of maximal sensitivity: one at the animal pole and the other at the vegetal pole. The damage done by noxious influences may result in actual breakdown of cells in a region of maximal sensitivity and may also lead to a depression of the developmental potential of the cells. Thus, the graded distribution of certain physiological properties appears to play a part in morphogenetic processes: physiological gradients are in fact also morphogenetic gradients.

Gradients in the embryo can be used to control development to a certain extent, by exposing the embryo to influences that, while reaching all parts, have a local effect as the result of differences in sensitivity. Disturbances of normal development often are the result of disruptions of gradients.

Organogenesis and Histogenesis

The primary organ rudiments continue to give rise to the rudiments of the various organs of the fully developed animal in a process called organogenesis. The formation of organs, even those of diverse function, shares some common features. As the organs form, so do their component tissues, in a process termed histogenesis.

A germinal layer, as the name implies, is a sheet of cells. An organ rudiment may be formed and separated from such a sheet in several ways. A groove, or fold, may appear within the layer, become closed into a tube, and then separated from the original layer. A tube once formed may be subdivided into sections by constrictions and dilations of the tube at certain points.

Alternatively, the germinal layer may produce a round depression, or pocket. The pocket may then separate from the layer as a vesicle, or it may elongate and branch at the tip while still connected with the layer. The latter method is common in the development of various glands and also the lungs in vertebrates.

Still another method of rudiment formation in a germinal layer is by the development of local thickenings, elongated or round, and detachment from the epithelial sheet. If a lumen appears later within such a body, the result may be the same as that achieved by folding—that is, a tube or vesicle may be formed. Indeed, the same sort of organ may develop even in related animals in either of these ways. The epithelial layer may further be cut up into segments, with the layer losing continuity, as in the formation of somites in vertebrates or similar mesodermal blocks in segmented invertebrates (e.g., annelids and arthropods).

Lastly, the cells of a germinal layer may give up their connection to each other and become a mass of loose, freely moving cells called embryonic mesenchyme. This mass gives rise to various forms of connective tissue but may also condense into more solid structures, including parts of the skeleton and the muscles.

Many organs are comprised of all three germinal layers. It is very common for glands, for instance, to derive their lining from an ectodermal or endodermal epithelium and their connective tissue (sometimes in the form of a capsule) from mesenchyme of mesodermal origin. Parts of ectoderm and endoderm cooperate also in the development of the lining of the alimentary canal, and mesoderm provides the connective tissue and muscular sheath of the canal.

Ectodermal Derivatives

The Nervous System

The vertebrate nervous system develops from the neural plate—a thickened dorsal portion of the ectoderm—which forms a tube. From the very start the tube is wider anteriorly, the end that gives rise to the brain. The posterior part of the neural tube, which gives rise to the spinal cord, is narrower and stretches as the embryo lengthens. Stretching involves the head to only a very minor degree.

The Brain and Spinal Cord

Constrictions soon appear in the brain region of the neural tube, subdividing it into three parts, or brain vesicles, which undergo further transformations in the course of development. The most anterior of the primary brain vesicles, called the prosencephalon, gives rise to parts of the brain and the eye rudiments. The latter appear in a very early stage of development as lateral protrusions from the wall of the neural tube, which are constricted off from the remainder of the brain rudiment as the optic vesicles. The rest of the prosencephalon constricts further into two portions, an anterior one, or telencephalon, and a posterior one, or diencephalon. The telencephalon gives rise, in lower vertebrates, to the smell, or olfactory, centre; in higher vertebrates and man, it becomes the centre of mental activities. The diencephalon, with which the eye vesicles are connected, was presumably originally an optic centre, but it has acquired, in the course of evolution, a function of hormonal regulation. The floor of the diencephalon forms a funnel-shaped depression, the infundibulum, which becomes connected with the pituitary, or hypophysis, the most important gland of internal secretion (*i.e.,* endocrine gland) in vertebrates. Indeed, the posterior lobe of the hypophysis is actually derived from the floor of the diencephalon. Tissues of the infundibulum and the posterior lobe of the hypophysis produce certain hormones (oxytocin and vasopressin) and stimulate the production and release of other hormones from the anterior lobe of the hypophysis.

The second primary brain vesicle, the mesencephalon, gives rise to the midbrain, which, in higher vertebrates, takes part in coordinating visual and auditory stimuli.

The third primary brain vesicle, the rhombencephalon, is more elongated than the first two; it produces the metencephalon, which gives rise to the cerebellum with its hemispheres, and the myelencephalon, which becomes the medulla oblongata. The cerebellum acts as a balance and coordinating centre, and the medulla controls functions such as respiratory movements.

The cells constituting the wall of the neural tube and, later, of the brain and spinal cord become arranged in such a way that they point into the central cavity of the tube. The differentiation of nervous tissue involves many cells abandoning their connection to the inner surface of the neural tube and migrating outward, where they accumulate as a mantle. The first cells to migrate become the neurons, or nerve cells. They produce outgrowths called axons and dendrites, by which the cells of the nervous system establish communication with one another to form a functional network. Some of the outgrowths extend beyond the confines of the brain and spinal cord as components of nerves; they establish contact with peripheral organs, which thus fall under the control of the nervous system. Cells migrating from the inner surface of the neural tube later in development become astrocytes, which are the supporting elements of nerve tissue.

The fate of nerve cells is dependent largely on whether they succeed, directly or indirectly (through other neurons), in connecting with peripheral organs. Nerve cells that fail to establish connections die. Thus, if in early stages of embryonic development, some organ, a limb rudiment for instance, is surgically removed, the nerve cells in the centres supplying nerves to such an organ are reduced in number, and the corresponding nerves also diminish or disappear. On the other hand, if an organ is introduced by transplantation into a developing embryo, the organ will be supplied by nerves from a nerve centre in which the number of cells apparently increases; no additional cells are provided, but cells that would otherwise have degenerated remain active and differentiate into functional neurons, thus satisfying the demand created by the additional organ.

Nerves do not consist entirely of outgrowths of neurons located in the brain and spinal cord. Many components of nerves are outgrowths of neurons, the cell bodies of which are located in masses called ganglia; there are three main types of ganglia: spinal ganglia, cranial ganglia, and ganglia of the autonomous nervous system. The spinal ganglia are derived from cells of the neural crest—the loose mesenchyme-like tissue that remains between the neural tube and skin after separation of the two. Part of the cells of the neural crest in the region of the trunk and tail accumulates in segmental groups (corresponding to the mesodermal somites) and provide fibres to peripheral organs and to the spinal cord. These fibres constitute the sensory pathways in the spinal nerves. The motor components of the spinal nerves—fibres that activate muscles—are outgrowths of neurons lying in the spinal cord. The ganglia of the cranial nerves are produced only in part from cells of the neural crest; an additional component comes from the epidermis on the side of the head. Cells of the epidermal thickenings called placodes detach themselves and contribute to the formation of the cranial ganglia and thus of the cranial nerves.

The ganglia of the autonomous (sympathetic) nervous system are derived, as are the spinal ganglia, from neural-crest cells, but, in this case, the cells migrate downward to form groups near the dorsal aorta, near the intestine, and even in the intestinal wall itself. The outgrowths of cells in these ganglia are the nerve fibres of the sympathetic nerves.

Major Sense Organs

The Eye

As has been pointed out, the rudiments of the eyes develop from optic vesicles, each of which remains connected to the brain by an eye stalk, which later serves as the pathway for the optic nerve. The optic vesicles extend laterally until they reach the skin, whereupon the outer surface caves in so that the vesicle becomes a double-walled optic cup. The thick inner layer of the optic cup gives rise to the sensory retina of the eye; the thinner outer layer becomes the pigment coat of the retina. The opening of the optic cup, wide at first, gradually becomes constricted to form the pupil, and the edges of the cup surrounding the pupil differentiate as the iris. The refractive systems of the eye and, in particular, the lens of the eye are derived not from the cup but from the epidermis overlying the eye rudiment. When the optic vesicle touches the epidermis and caves in to produce the optic cup, the epidermis opposite the opening thickens and produces a spherical lens rudiment. The lens develops by an induction by the optic vesicle on the epidermis with which it comes in contact. A further influence emanating from the eye changes the epidermis remaining in place over the lens into a transparent area, the cornea. Influence of the optic cup on the surrounding mesenchyme causes the latter to produce a vascular layer around the retina and, outside of that, a tough fibrous or (in some animals) even a partly bony capsule called the sclera. Thus a complex interdependence of different materials produces the fully developed and functional vertebrate eye.

The Ear

The main part of the ear rudiment is derived from thickened epidermis adjoining the medulla. This area of the epidermis invaginates to produce the ear vesicle, which separates from the epidermis but remains closely apposed to the medulla. The ear vesicle becomes complexly folded to produce the labyrinth of the ear. Subsequently, a group of cells of the ear vesicle becomes detached and gives rise to the acoustic ganglion. Neurons of this ganglion become connected by their nerve fibres to the sensory cells in the labyrinth, on the one hand, and with the brain (the medulla), on the other. The ear vesicle, acting on the surrounding mesenchyme, induces the latter to aggregate around the labyrinth and form the ear capsule. Further parts with various origins are added to the ear: the middle ear, from a pharyngeal pouch and the associated skeleton, and the external ear (where present), from epidermis and dermis.

The Olfactory Organ

The olfactory organ develops from a thickening of the epidermis adjacent to the neural fold at the anterior end of the neural plate. This thickening is converted into a pocket or sac but does not lose connection with the exterior. The openings of the sac become the external nares, and the cavity of the sac becomes the nasal cavity. Some cells of the olfactory sac differentiate as sensory epithelium and produce nerve fibres entering the forebrain. In most fishes the olfactory sac does not communicate with the oral cavity; in lungfishes and in terrestrial vertebrates, however, canals develop from the olfactory sacs to the oral cavity, where they open by internal nares. A cartilaginous capsule forms around the olfactory organ from cells believed to have been derived from the walls of the sac itself, and thus it is ectodermal in origin.

Gustatory and other Organs

Gustatory organs in the form of taste buds develop as local differentiations of the lining of the oral cavity but also, in fishes, in the skin epidermis. They are supplied with nerve endings, as are several other sensory bodies scattered among the tissues and organs of the developing body.

The Epidermis and its Outgrowths

The major part of the ectodermal epithelium covering the body gives rise to the epidermis of the skin. In fishes and aquatic larvae of amphibians, the many-layered epidermis is provided with unicellular mucous glands. In terrestrial vertebrates, however, the epidermis becomes keratinized; i.e., the outer layers of cells produce keratin, a protein that is hardened and is impermeable to water. During the process of keratinization, many cell components degenerate and the cells die; the layer of keratinized cells is therefore shed from time to time. In reptiles the shedding may take the form of a molt in which the animal literally crawls out of its own skin. It is less well known that frogs and toads also molt, shedding the surface keratinized layer of their skin (which is usually eaten by the animal). In birds and mammals, keratinized cells are shed in pieces that are sloughed off, rather than in extensive layers. In many vertebrates local thickenings of the keratinized layer appear in the form of claws, hooves, nails, and horns.

The epidermis is only the superficial layer of the skin, which is reinforced by the dermis, a connective tissue layer of a much greater thickness. The cells of the dermis are derived from mesoderm and neural-crest cells. In particular the pigment cells found in the dermis of fishes, amphibians, and reptiles are of neural-crest origin. The pigment in the skin of birds and mammals (and also in hairs and feathers) is also produced by neural-crest cells, but in these animals the pigment cells penetrate into the epidermis or deposit their pigment granules there.

The structure of the skin is further complicated by the development of hairs and feathers, on the one hand, and of skin glands, on the other. Hairs and feathers develop from a somewhat similar kind of rudiment. The development starts with a local thickening of the epidermal layer, beneath which a group of mesenchyme cells accumulate. In the case of hairs, the epidermal thickening proliferates downward and forms the root of the hair, from which the shaft then grows outward, emerging on the surface of the skin. In the case of feathers, the epidermal thickening bulges outward to form a hollow finger like protrusion with a connective tissue core. Secondarily, the shaft of the feather branches characteristically to produce barbs and barbules. In both cases, however, the final structure—shaft of the hair and shaft barbs and barbules of the feather—consists of keratinized and, thus, dead cells.

The skin of amphibians and mammals (but not of birds and reptiles) is provided with numerous skin glands, which develop as ingrowths from the epidermis. A peculiar type of skin gland is the mammary gland of placental mammals. In the first stage of development, mammary-gland rudiments resemble hair rudiments; they are thickenings of the epidermis, with condensed mesenchyme on their inner surfaces. In some mammals (rabbit, man) two continuous epidermal thickenings called mammary lines stretch along either side of the belly of the embryo. Parts of the line corresponding in number and position to the future glands enlarge while the rest of the thickening disappears. The initial thickenings proliferate inward and produce a system of ramified cords, solid at first but hollowed out later, which become the lactiferous, or milk-bearing, ducts of the gland. Further branching at the tips of the ducts gives rise to smaller ducts and to the secretory end sacs, or alveoli, of the gland.

Mesodermal Derivatives

The Body Muscles and Axial Skeleton

The somites, formed in the early stages of development from the upper edges of the mesodermal mantle adjoining the notochord, are complex rudiments that subdivide and give rise to very diverse body structures. The coelomic cavity, present initially, becomes obliterated by the side-to-side flattening of the somites, so that the thinner, outer parietal layer of the somite comes in close contact with its thicker visceral layer. The visceral layer of the somite very early subdivides into two parts. The upper, dorsolateral part called the myotome remains compact, giving rise to the body muscles. The lower, medioventral part of the somite, called the sclerotome, breaks up into mesenchyme, which contributes to the axial skeleton of the embryo—that is, the vertebral column, ribs, and much of the skull. The parietal layer of the somite, at a later stage, is converted into mesenchyme that, together with components of the neural crest, gives rise to the dermis of the skin and, for this reason, is called the dermatome.

The cells of the myotome are elongated in a longitudinal direction and become differentiated as muscle fibres. The myotomes, originally situated dorsally, expand on either side, penetrating between the skin on the outside and the lateral plates of the mesoderm on the inside, until they meet midventrally; the whole body is thus enclosed in a layer of developing muscle. As the somites and myotomes are segmented, so are the muscles derived from them. Metamerism, or segmentation, a feature in the embryos of all vertebrates, remains preserved only in the adults of fishes and of terrestrial vertebrates that have elongated bodies (salamanders, snakes); it becomes largely erased in four-footed animals that depend on their limbs for locomotion.

The mesenchyme derived from the sclerotomes condenses as cartilage around the notochord and the spinal cord. It forms the cartilaginous vertebral column and ribs. In the head region it produces a part of the cartilaginous skull, mainly its posterior and ventral parts; anteriorly the somitic mesenchyme is supplemented by mesenchyme from the neural crest. Cartilaginous capsules of the olfactory organ and the ear fuse with the cartilaginous capsule surrounding the brain; to this complex are also added cartilages associated with the jaws and gill skeleton. Cartilage in the vertebral column and in the skull is replaced later in the bony fishes and in the terrestrial vertebrates by bone. At a still later stage, dermal bones are added, which, while they have no precursors in the cartilaginous skeleton, develop in the adjoining mesenchyme.

The Appendages: Tail and Limbs

The tail in vertebrates is a prolongation of the body beyond the anus. It develops in early stages from the tail bud, immediately dorsal to the blastopore. Material for the tip of the tail is situated slightly forward from the edge of the blastopore. The elongation of the back of the body is greater than that of the belly; as a result the tip of the tail bud is carried beyond the blastopore and thus beyond the anus, which, in the developed embryo, marks the position of the blastopore. The consequence is that a section of the dorsal surface of the embryo comes to lie on the ventral surface of the tail; i.e., becomes inflected. The tail bud is formed from parts that have already been differentiated to a certain extent; prolongations of the neural tube and of the notochord are involved, and endoderm extends into the tail rudiment as the postanal gut, which, however, soon degenerates. The bud is also encased in ectodermal epidermis. In amphibians the somites of the tail are not derived from the chordamesodermal mantle but from the inflected posterior portion of the neural plate, which loses its nervous nature and becomes subdivided into segments corresponding to the somites of the trunk. In higher vertebrates the cells in the interior of the tail bud have an undifferentiated appearance and form a growth zone, at the expense of which parts of the tail (neural tube, notochord, somites) are extended backward as the tail elongates.

The paired limbs of vertebrates derive their first rudiments from the upper edge of the lateral plate mesoderm. The parietal layer becomes thickened, and cells escape from the epithelial arrangement and form a mesenchymal mass adjoining the ectodermal epithelium at the surface of the body. The ectodermal epithelium over the mass of mesenchyme likewise becomes thickened. In higher vertebrates, the accumulation of mesodermal cells and the thickening of the epidermis occur along the entire length of the trunk, from neck to anus, but in the middle of the trunk they soon disappear, and only the most anterior and the most posterior sections develop further into the rudiments of the forelimbs and hindlimbs, respectively. In fishes, the rudiments of the pectoral and pelvic fins are more extended anteroposteriorly in earlier than in final stages.

The mesodermal masses of the limb rudiments proliferate, and, covered with thickened epidermis, form on the surface of the body conical protrusions called the limb buds, which, once formed, possess all the materials necessary for limb development. Limb buds may be transplanted into various positions on the body or on the head and there develop into clearly recognizable limbs, conforming to their origin, whether a forelimb or hindlimb, a wing or a leg in birds. This specificity of the limb is carried by the mesodermal part of the rudiment, but a complex interaction between the mesodermal mesenchyme and the ectodermal epidermis is necessary for the normal development of the limb. In four-limbed vertebrates (tetrapods), the tips of the limb buds become flattened and broadened into hand or foot plates. The edge of the plate is indented, forming the rudiments of the digits. Meanwhile, local areas of the mesodermal mesenchyme in the interior of the limb rudiment condense; these are the rudiments of the various components of the limb skeleton. In fishes, small outgrowths from the myotomes enter the limb rudiment to form the muscles of the fins. In tetrapods, however, the limb muscles develop from the same mass of mesenchyme that gives rise to the skeleton. Thus the muscles of the body and the muscles of the limbs have different origins—the first develop from the myotomes (thus from the somites), and the second develop from the lateral plate mesoderm via the limb buds.

The nerves supplying the limbs grow into the limb rudiments from the spinal cord and the spinal ganglia. The nerves are guided in some way by the limb rudiments, for, if limb rudiments are displaced by transplantation to an abnormal position, the nerves still find their way and establish normal relationships to the limb muscles. Limb rudiments transplanted to sites very far from their normal positions induce local nerves to enter the limb, thereby making it motile.

Excretory Organs

The kidneys of vertebrates consist of a mass of tubules that develop from the stalks of somites called nephrotomes. In some primitive vertebrates such as cyclostomes, the nephrotome in each segment gives rise to only one tubule, but, in the great majority of vertebrates, mesenchyme from adjacent nephrotomes fuses into a common mass that differentiates into a number of nephric tubules irrespective of the original segmentation of the mesoderm. Under primitive conditions each tubule opens by a funnel (the nephrostome) into the coelomic cavity; the opposite ends of the tubules fuse to form the collecting ducts of the kidney. A collection of capillaries (the glomerulus) becomes associated with the nephric tubule, forming its filtration apparatus. The glomerulus may be situated in the coelomic cavity opposite the nephrostome or, in all the more advanced animals, intercalated into the nephric tubule, forming with the latter a renal corpuscle of the kidney. In adults of all vertebrates above the amphibians, the nephrostomes disappear (or are never formed), so that the tubule begins with the renal corpuscle. Parts of the kidney in vertebrates can be distinguished as the pronephros (most anteriorly, at the forelimb level), the mesonephros (in the midtrunk region), and the metanephros (in the pelvic region). The three sections of the kidney develop at different stages, starting with the pronephros and ending with the metanephros. In their morphology and mode of development, the anterior parts show more primitive conditions than the posterior ones. The pronephros, developing early in embryo formation, is the functional kidney of fish and amphibian larvae. Its collecting duct opens into the hindmost part of the intestine, called the cloaca, and later also serves as the collecting duct of the mesonephros. In reptiles, birds, and mammals, the pronephros is nonfunctional, although even in these animals its duct persists as the mesonephric duct. The mesonephros develops later and replaces the pronephros as the functional kidney of adult fishes and amphibians and of the embryos of reptiles, birds, and mammals. The tubules of the mesonephros link up with the duct derived from the pronephros. The pronephric duct in fact stimulates the development of mesonephric tubules, and, in its absence, the mesonephros does not develop at all.

The metanephros is found only in reptiles, birds, and mammals. It replaces the mesonephros of the early embryonic stages and continues as the functional kidney in the postembryonic and adult life of these animals. The metanephros develops from mesenchyme derived from the nephrotomes of the posterior part of the trunk and lying dorsal to the mesonephric duct. The actual differentiation is initiated by a dorsal outgrowth of the mesonephric duct, called the ureteric bud. The ureteric bud grows in the direction of the mesenchyme and becomes the ureter. Having penetrated the mass of mesenchyme, it starts to branch, producing the collecting tubules of the kidney; the mesenchyme, meanwhile, in response to the influence of the duct and its branches, aggregates to form the excretory tubules of the kidney. The influence of the ureter is indispensable for the development of the metanephric excretory tubules, for, if the ureter fails to develop or, in its outgrowth, stops short of reaching the kidney-producing mesenchyme, no kidney develops.

Circulatory Organs

The rudiment of the heart in vertebrates develops from the ventral edges of the mesodermal mantle in the anterior part of the body, immediately adjoining the pharyngeal region. A group of mesodermal cells breaks away from the ventral edge of the lateral plate, takes a position just underneath the pharyngeal endoderm, and becomes arranged in the form of a thin-walled tube, which will become the endocardium, or lining of the heart. In vertebrates with complete cleavage, the endocardial tube is single and medial from its start. In higher vertebrates with meroblastic cleavage—reptiles, birds, and mammals—the embryo in early stages of development is flattened out on the surface of the yolk sac; therefore, what are morphologically the ventral edges of the mesodermal mantle lie far apart on the perimeter of the blastodisc. As a result of this arrangement, two endocardial tubes are formed, one on either side of the embryo. Subsequently, when the embryo becomes separated from the yolk sac, the two endocardial tubes meet in the midline ventral to the pharynx and fuse, producing a single heart rudiment. After the formation of the endocardium, or the lining of the heart, the coelomic cavity in the lateral plate mesoderm adjoining the heart rudiment expands slightly and envelops the endocardial tube or tubes. The heart muscle layer, or myocardium, develops from the visceral (splanchnic) layer of the lateral plate that is in contact with the endocardial tube; the parietal (somatic) layer of the lateral plate forms the pericardium, or covering of the heart. The portion of the coelom surrounding the heart becomes separated from the rest of the body cavity and develops into the pericardial cavity.

The endocardial tube branches anteriorly into two tubes, the ventral aortas; a similar branching of the endocardial tube posteriorly forms the two vitelline veins, which carry blood from the midgut endoderm or from the yolk sac (when present) to the heart.

In its earliest development, the heart rudiment shows a degree of dependence on the adjoining endoderm. The whole of the endoderm can be removed in newt embryos in the neural-tube stage. In such endodermless embryos, the heart fails to develop, even though the mesoderm destined to form the heart rudiment is left intact.

The heart is initially a straight tube stretching in an anteroposterior direction. Rather early in development, however, it becomes twisted in a characteristic way and subdivided into four main parts: the most posterior, the sinus venosus; the atrium, which comes to lie at the anteriorly directed bend of the tube; the ventricle, occupying the apex of the posteroventrally directed inflexion; and, most anteriorly, the conus arteriosus. In the course of development in the more advanced vertebrates, the atrium and ventricle become partially or completely subdivided into right and left halves. In amphibians, only the atrium is separated into two halves, by a partition starting from the posterior end. In reptiles, a partition separates the atria and part of the ventricle. In birds and mammals, the subdivision of the heart is complete, with two atria and two ventricles.

The complete subdivision of the heart is important for separating the pulmonary, or lung, blood supply from the general body circulation. But, if this separation developed early in the embryo, it would create difficulties, since the lungs of the embryo are not functional; the enrichment of the blood with oxygen occurs instead in the placenta. The partition between the atria in mammalian embryos remains incomplete, so that blood returning from the body and from the placenta enters into the right half of the heart but is shunted (through the interatrial foramen) into the left half of the heart and thence again into general circulation. At birth, however, the interatrial foramen is closed by a membraneous flap, and oxygen-depleted blood from the body enters the right atrium, is channelled into the right ventricle, and thence to the lungs for oxygenating.

In an adult vertebrate, blood vessels extend to all parts of the body. It would seem that channels for the supply of blood are provided in proportion to the local demand of the tissues; progressively developing organs or parts with particularly intensified function always receive an increased blood supply. The rudiments of blood vessels are always aggregations of mesenchyme cells. In any blood vessel the endothelial tube is formed first, and the muscular and elastic layers are added later.

The main blood channels in vertebrates develop in certain favoured situations; namely (1) between the endoderm and lateral plate mesoderm; (2) around the kidneys, especially the pronephros and mesonephros; and (3) in connection with the heart, which is a special case of the first category.

From the paired forward extensions from the heart, the ventral aortas, loops develop between the pharyngeal clefts. These are the aortic arches, which served originally to supply blood to the gills in aquatic vertebrates. The arches are laid down in all vertebrates, six or more being found in cyclostomes and fishes; six are present in the embryos of tetrapods, but the first two are degenerate. The arches of the third pair develop as the carotid arteries, supplying blood to the head. Those of the fourth pair (and, exceptionally, in urodeles also the fifth) join dorsally to form the dorsal aorta, providing blood to most of the body. These are the systemic arches. The arches of the sixth pair are the pulmonary arches; in embryos they carry blood to the dorsal aorta, as well as to the lungs, but in fully developed amniotes (reptiles, birds, and mammals), they carry blood only to the lungs.

The paired posterior extensions of the heart of the early embryo are the vitelline veins, whose branches spread out between the lateral plate mesoderm and the endoderm, especially the endoderm of the yolk sac, when present. On their way to the heart, the vitelline veins pass through the liver and break up into a system of small channels—the hepatic sinusoids. Parts of the vitelline veins lying posterior to the liver become the hepatic portal veins, which carry blood from the intestine to the liver; the parts of the vitelline veins anterior to the liver become the hepatic veins, which carry blood from the liver to the sinus venosus in lower vertebrates (anamniotes), but become the anterior section of the postcaval vein in amniotes.

Whereas the vitelline veins and, later, the hepatic portal vein carry blood from the endodermal parts of the embryo and from the yolk sac to the heart, the blood from the mesodermal and ectodermal parts is returned to the heart through a system of cardinal veins. These latter veins start their development in the form of an irregular sinus around the pronephros, connected by the common cardinal veins (ducts of Cuvier), on either side, to the sinus venosus. Extensions anteriorly and posteriorly give rise to the precardinal and postcardinal veins, respectively. The postcaval vein, present in terrestrial vertebrates, is a late acquisition, both in evolution and in embryogenesis; it is a result of the intercommunication of several venous channels, including the anterior portion of the vitelline veins.

The first blood cells in vertebrate embryos form in association with the intestinal endoderm on the yolk sac. Groups of mesoderm cells derived from the splanchnic layer of the lateral plate (extra-embryonically in cases in which a yolk sac is present) become so-called blood islands, which are particularly conspicuous on the yolk sac of bird embryos (in the area vasculosa). In bird's eggs, the internal cells of the blood island start producing hemoglobin (gas-carrying component of blood) and become the first red blood cells (erythrocytes) as early as the second day of incubation. The outer cells of the blood islands develop into an endothelial layer and form a network of blood vessels covering part of the surface of the yolk sac. The network acquires a connection to the vitelline veins and vitelline arteries (the latter being branches of the dorsal aorta); thus the blood corpuscles formed in the blood islands can enter the general blood circulation.

At later stages of embryogenesis, blood-cell formation shifts from the blood islands to the liver and, still later, to the bone marrow. The lymphatic system, in a manner similar to the blood vessels, develops by the local aggregation of connective tissue to form lymphatic vessels.

Reproductive Organs

In considering the development of reproductive organs, distinctions must be made between: (1) the origin of sex cells (gametes), (2) the origin and differentiation of the sex glands, or gonads (ovaries and testes), and (3) the origin and development of the supporting parts of the reproductive system (*e.g.*, genital ducts, copulatory organs).

The germ (germinal) cells, which eventually give rise to the gametes, are often segregated from the somatic, or body, cells at a very early stage—during cleavage and before the subdivision of the embryo into ectoderm, mesoderm, and endoderm. In the invertebrate nematodes, the very first of these primordial germ cells is identifiable after as few as five divisions of the egg cell. The germ cell retains the large chromosomes present in the fertilized egg; in the somatic cells the chromosomes become fragmented. Subsequently, the single germ cell gives rise, by mitotic divisions, to all the gametes in the gonad.

In vertebrates, primordial germ cells arise outside the gonads, but they cannot be distinguished in early cleavage stages. In amphibians, cytoplasm at the vegetal pole, rich in ribonucleic acids, becomes incorporated into a number of cells, which, during cleavage and gastrulation, lie among the yolky endoderm cells. Later they migrate into the mesodermal layer and become incorporated into the rudiments of the gonads. In higher vertebrates, primordial germ cells can be recognized in the extra-embryonic endoderm of the yolk sac. In mammals, these cells subsequently migrate into the mesoderm and are located in the gonad rudiments. The mouse embryo, for example, originally has fewer than 100 primary germ cells; during their migration, however, their numbers increase as a result of repeated divisions, to 5,000 or more in the gonads.

Although the primordial germ cells either may appear before the separation of germinal layers or be found originally in the endoderm, the gonads are invariably of mesodermal origin. In vertebrates, the first trace of gonad development is a thickening of the coelomic lining on either side of the dorsal mesentery and medial to the kidney rudiments. The thickening, elongated anteroposteriorly, is known as the germinal ridge. The ridge protrudes into the coelomic cavity, and the fold of thickened epithelium becomes filled with mesenchyme. At this stage the primordial germ cells invade the rudiments of the gonads and become associated with the somatic cells of the germinal ridge. In the functionally differentiated gonads, only the actual gametes and their predecessors (spermatogonia and oogonia) are derived from the primary germ cells; the supporting cells are somatic cells of local mesodermal origin. In the ovaries, the follicle cells surrounding and nourishing the young egg cells (oocytes) are of somatic origin, as are also the connective tissue and blood vessels of the gonad. In the testes, supporting elements called Sertoli cells are somatic cells, as are the interstitial cells, which are scattered between the sperm-carrying tubules of the testes and believed to be the source of male hormones.

In the early stages of their development—even while the gonad rudiment is being invaded by primordial germ cells—the female and male gonads are in an indifferent stage. Only later does tissue differentiation of the gonads begin and male or female gonadal development proceeds.

The genital ducts, by which the eggs and sperm are carried away from the gonads, are, in vertebrates, linked with the excretory system. In the male, the seminiferous tubules connect with the nephric tubules of the mesonephros, and the sperm are carried to the exterior by way of the mesonephric duct. In males of lower vertebrates, the mesonephric duct thus serves as a channel both for urine and for sex cells. In amniotes the development of the metanephros as the urine excreting organ has freed the mesonephric duct to carry products associated only with reproduction. In the female, a separate duct, the paramesonephric duct (Müllerian duct), develops beside the mesonephric duct. At its anterior end it utilizes the funnels of the pronephric tubules as its entrance (ostium). The paramesonephric duct develops initially in both female and male embryos. The ducts remain in an indifferent stage longer than the gonads. Eventually the sex hormones produced by the differentiating gonads cause a corresponding differentiation of the ducts. The mesonephric ducts, which become reduced in female embryos, remain in male embryos as ducts for conveying sperm (ductus deferens). The paramesonephric ducts, on the other hand, degenerate in male embryos but become the oviducts in female embryos. In mammals, the terminal portions of the paired oviducts differentiate as two uteri, which, in primates and man, fuse to form a single uterus.

In all terrestrial vertebrates except the placental mammals, the genital ducts, as well as the ducts of excretory organs, open into the cloaca. In mammals, however, the cloaca becomes subdivided into a dorsal part, which conveys the feces, and a ventral part, which receives excretory and genital products. In male mammals the excretory and genital ducts remain connected, having the urethra as their common outlet; in females the urethra serves only for the passage of urine and the uterus opens separately by means of the vagina. In nearly all vertebrates, the male nephric duct is utilized in some degree for the conduction of sperm.

Copulatory organs have developed independently in several groups of vertebrates having internal fertilization. The penis in mammals develops from an outgrowth called the genital tubercle, located at the anterior edge of the urinogenital orifice. The tubercle is laid down in a similar way in embryos of both sexes, and the region of the urinogenital orifice remains in an indifferent state even longer than do the genital ducts. In a comparatively late stage of embryonic life the genital tubercle of male embryos encloses the urethral canal and becomes the penis; in female embryos it remains small and becomes the clitoris.

Endodermal Derivatives

The Alimentary Canal

The alimentary canal is the chief organ developing from endoderm. The way it forms depends on the type of egg cleavage. In eggs with holoblastic (complete) cleavage, after gastrulation the invaginated mass of endoderm lines the archenteron, the cavity of which becomes the alimentary canal, or gut. In eggs with meroblastic (partial) cleavage—and also in mammals (despite their complete cleavage)—the endoderm is produced in the form of a sheet lying flat over the yolk-sac cavity. Subsequently, folds of endoderm and splanchnic mesoderm appear—first anteriorly, then laterally, and lastly posteriorly—and sink, converging ventrally under the embryo and cutting off the future gut cavity from the cavity of the yolk sac. The most anterior and posterior portions of the gut separate, but the middle part remains in open communication with the yolk sac throughout embryonic life, eventually becoming reduced to the yolk stalk, which passes through the umbilical cord.

The alimentary canal of vertebrates becomes differentiated into the oral cavity, pharynx, esophagus, stomach, and intestine. Whether derived from an archenteron or formed by folding of the endodermal sheet, the canal initially does not possess an opening at its anterior end. This is also the case in some lower chordates and echinoderms, which are grouped together with vertebrates as the Deuterostomia, or animals with secondary mouths.

In vertebrates, a mouth forms by a rupture at the anterior end, where the endoderm is in contact with ectoderm. The ectoderm of the future mouth region becomes depressed, forming a mouth invagination, or stomodaeum. The ectodermal and endodermal layers separating the cavity of the stomodaeum from the gut fuse to form the oropharyngeal membrane, which thins and ruptures, providing free passage from the exterior to the gut. Because of its mode of origin, the oral cavity is in part lined by ectoderm and in part by endoderm, the two parts becoming indistinguishable. Before the oropharyngeal membrane ruptures, however, a small pocket forms on the dorsal side of the stomodaeal invagination. This, the rudiment of the anterior lobe of the hypophysis, becomes apposed to the ventral surface of the diencephalon and loses its connection with the mouth cavity.

The anal opening in some exceptional cases (urodele amphibians) is derived directly from the blastopore, which persists as a narrow canal after completion of gastrulation. In other vertebrates, however, the anus develops either near the location of the former blastopore or in a corresponding region at the posterior end of the embryo, where the last remnants of mesoderm migrated to the interior. It is thus claimed that the anus in vertebrates is derived, directly or indirectly, from the blastopore. The mode of formation of the opening is somewhat similar to that of the mouth. A slight invagination of the ectoderm occurs, and a cloacal membrane forms, separating the ectodermal invagination from the gut cavity. The membrane ruptures later to provide the anus.

The Pharynx and its Outgrowths

The anterior portion of the endodermal gut, lying immediately posterior to the mouth cavity, expands laterally as the pharynx. The lateral pockets of the pharyngeal cavity, called the pharyngeal pouches, perforate the mesodermal layer, reach the ectoderm, and break through to form pharyngeal, or gill, clefts. In fishes and larvae of amphibians, these clefts develop gills and become respiratory organs. Pharyngeal pouches develop in the early embryos of all vertebrates, including the air-breathing terrestrial reptiles, birds, and mammals. The number of pouches has been reduced in the course of evolution from six or more to four in tetrapods, and the posterior pouches may not actually break through.

The consistent development of pharyngeal pouches and clefts indicates their importance in vertebrate development. Many parts of the vertebrate body are derived from, or dependent on, the pharyngeal pouches; for example, the aortic arches—the most important blood vessels of a vertebrate—develop between successive pharyngeal pouches. Skeletal visceral arches also occur between consecutive pharyngeal pouches (they do not develop if the pharyngeal pouches are prevented from developing). In adult terrestrial vertebrates, parts of the visceral arches are transformed into the hyoid apparatus, supporting the tongue, the auditory ossicles, and parts of the larynx and trachea. Furthermore, some of the material of the pharyngeal pouches is utilized for the formation of the parathyroid glands and the thymus; the former are indispensable glands of internal secretion, and the latter are a source, in mammals, of cells that produce antibodies. The pharynx also produces the rudiment of the thyroid gland as a ventral outgrowth.

The Liver, Pancreas and Lungs

Three additional important organs develop from the endoderm: the liver, the pancreas, and the lungs. The liver develops as a ventral outgrowth of the endodermal gut just posterior to the section that eventually will become the stomach. Initially, the liver takes the form of a tubular gland, but it soon acquires a close relationship to the blood sinuses and capillaries, forming lobules around blood vessels rather than around glandular ducts. The pancreas develops from three independent rudiments: two ventral ones, formed just posterior to the liver rudiment, and a dorsal one. The ventral and the dorsal rudiments fuse in most vertebrates to form one organ with a complicated system of ducts opening into the duodenum, a portion of the small intestine. The lungs develop from a ventral hollow outgrowth of the gut, which is located just posterior to the pharyngeal region; the outgrowth branches into a right and left trunk that grow posteriorly beside the esophagus and then expand into hollow sacs, in lower terrestrial vertebrates, or into a system of tubes, in birds and mammals.

The endodermal parts of the alimentary system are, along their entire length, encased by the splanchnic mesoderm of the lateral plates. The coelomic cavities of the right and left sides fuse ventral to the gut but remain separated dorsally by their respective walls, which form the dorsal mesentery—a double membrane by which the gut is suspended from the dorsal side of the body cavity and through which blood vessels and nerves reach the gut. The layer of splanchnic mesoderm next to the endoderm produces the connective tissue and muscular layers of the gut. During development of the glands of the alimentary canal (e.g., pancreas, salivary glands), the mesoderm forms a connective tissue capsule around the branching tubules of the gland. The development of the tubules is dependent on this mesodermal capsule and cannot proceed without it.

Postembryonic Development

After partially developing within the egg membranes or within the maternal body, the newly formed individual emerges. The new animal is then born (ejected from the mother's body) or hatched from the egg. The condition of the new organism at the time of birth or hatching differs in various groups of animals, and even among animals within a particular group. In sea urchins, for example, the embryo emerges soon after fertilization, in the blastula stage. Covered with cilia, the sea-urchin blastula swims in the water and proceeds with gastrulation. Frog embryos emerge from the egg membranes when the main organs have already begun to develop, but functional differentiation of the tissues is unfinished; for instance, the components of the eyes and ears are far from complete, the mouth is not yet open, and the gut is filled with yolk-laden cells. Certain birds (called precocial) emerge from the egg covered with downy feathers and can run about soon after hatching, whereas others (altricial) hatch naked, with only rudiments of feathers, and are quite unable to move around. Among mammals there is a great range in the degree of development at birth. In marsupials, such as opossums and kangaroos, the young are born incompletely developed and very small; the young are then kept for a long time in the pouch of the mother, all the while firmly attached to the teats and suckling. Many small mammals are helpless at birth. Mice are born naked and blind; puppies and kittens are born covered with fur but with unopened eyes. Newborn human babies have their eyes open but cannot move themselves about for several months. Hoofed mammals, on the other hand, bear young that can stand up and run after their mothers within a few hours of birth.

In birds the hard shell is broken by the hatchling's beak, which is provided with a sharp tubercle on its top. A similar "egg tooth" appears on the tip of the snout of hatchling reptiles. Many arthropods have a preformed line of fragility that allows part of the eggshell to be burst open like a lid, allowing the young to emerge. Birth in mammals is affected through the contraction of smooth muscles of the uterus.

The Larval Phase and Metamorphosis

The organism emerging from the egg or from the maternal body, apart from being incompletely developed, may have an organization more or less different from that of an adult. In some cases the difference is so great that, without knowing the origin of the eggs or without following the young through their full course of development, it would be impossible to know that the young and the adult are of the same animal species. Such young, called larvae, transform into the adult form by a process of metamorphosis. The term larva also applies to young that resemble the adult form but differ from it in some substantial respect, as in possessing organs not present in the adult or in lacking an important structure (apart from sex glands and associated parts, which tend to develop later in life in most animals). Larvae in different animals have special names given to them, such as the tadpole of frogs, the caterpillar of butterflies, and the fry of fishes.

The Larval Stage

The development of the embryo into a larva rather than directly into an organism similar to the adult has various advantages. At the time of emergence from the egg, the new individual is relatively small, and the organization that enables the adult to lead a particular mode of life may not be suitable for a miniature copy of the adult. The larva may have to procure food for it and, being small, may not be able to feed in the same way as the adult. It also may not be able to use effectively the same defense mechanisms the adult possesses. The larval stage enables an animal to avoid such hazards; it provides a mode of life and corresponding organization better suited to the smaller size of the newly emerged organism. Another advantage is that the larva may be able to exploit an entirely different environment because its organization is very different from that of the adults. A terrestrial adult may have aquatic larvae, a flying adult may have burrowing larvae, and a parasitic adult may have a free-living larva. A third advantage of a larval stage emerges in animals whose adult stages are sessile or restricted in their movements; the larvae can move freely, either of their own accord or on water currents. In this way the larvae of sedentary animals serve for the dispersal of the species. Lastly, the larval stage is of great advantage for certain internal parasites, which, once inside a host, cannot transfer to another. New hosts are infected instead by the larval stages. (The usual means of attaining this end is for the parasite to produce enormous quantities of eggs and rely on the passive entry of the eggs into the new host with food. A more efficient way, however, is for a mobile larva to enter the new host actively.)

A large number of marine invertebrates possess floating larvae that have hairlike projections (cilia) as their means of locomotion. There are three main types of larvae, characteristic of large subdivisions of the animal kingdom.

The planula larva of coelenterates has an elongated shape and cilia covering its entire surface. The internal organization is simple, hardly beyond differentiation into ectoderm and endoderm in the interior. The larva does not feed but serves only for dispersal.

The trochophore larva is found in many marine invertebrates. Typically, as in polychaetes, it has an alimentary canal with mouth and anus and a ring of ciliated cells arranged anterior to the mouth. It also possesses a sensory organ and rudiments of mesoderm. Cilia around the mouth bring in food—unicellular plants and other small particles. The larva thus not only serves for dispersal but also feeds and grows before it transforms into an adult worm. Other trochophore larvae are found in marine mollusks and in certain marine worms. The larva of echinoderms is similar to the trochophore in possessing a gut and a ciliary band, but the arrangement of the latter is different. The echinoderm larva also feeds and grows as well as serves for dispersal.

Larvae of very different kinds are found in many arthropods. In crustaceans the larva, called nauplius, does not differ substantially in mode of life or means of locomotion from the adult but has fewer appendages than the adult. A typical crustacean nauplius has three pairs of legs and an unpaired simple eye. Additional pairs of appendages and paired compound eyes appear in the course of a sometimes prolonged development. In insects the larva differs from the adult by the absence of wings but, in addition, may have a different mode of life and different way of feeding. Among chordates the tunicates (sea squirts) deserve attention; the larval form is a free-swimming creature, showing unmistakable relation to vertebrates, but the adult is sedentary, with much reduced nervous and muscular systems. The tadpole of a frog differs from the adult in being totally aquatic, in possessing a tail and gills for respiration, and in having a mouth adapted for feeding on plants. The adult frog is adapted to land life, except for reproductive periods, has no tail and no gills, and is an active predator.

Metamorphosis

Metamorphosis, the transformation of the larva into an adult, is a more or less complicated process depending on the degree of difference between the two forms. The transformation may be gradual, extend over a long period, and involve a number of intermediate stages; alternatively, the transformation may be achieved in one step. In the latter case, especially if the difference between the larva and adult is great, large parts of the body of the larva, including all the specifically larval organs, disintegrate (necrobiotic metamorphosis). At the same time, organs of the adult are built up, sometimes from reserve groups of cells that remain undifferentiated or nonfunctional in the larva. A good illustration of the distinction between gradual and abrupt metamorphosis occurs among the insects. In more primitive insects, such as cockroaches and grasshoppers, metamorphosis is gradual. The larva, often referred to as a nymph, has more or less the same organization as the adult, or imago; it feeds in a similar way but differs from the adults in lacking wings and in having incomplete sex organs. The wings appear in later stages of larval life; they are small at first but increase with each molt, and they attain full size and functional capacity at the last (imaginal) one. The larva of other insects, such as beetles, butterflies, and wasps, is a grub or caterpillar, a wormlike creature not even remotely resembling the adult. The difference in organization is so profound that the transformation cannot be achieved gradually, and an intermediate resting, or pupal, stage is interposed between the larva and imago. The pupa neither feeds nor moves, as the larval organs inside are destroyed and replaced with organs of the adult, including wings and sex organs. Eventually, when formation of the adult organs is complete, the pupal skin is cast off, and the adult emerges. The destruction of the larval parts may be far reaching and include even the skin and most of the alimentary canal. The tissues of the adult are formed from groups of reserve cells that were present all along in the larva as imaginal disks.

Necrobiotic metamorphosis is observed in the tunicate larva, in which the tail, including notochord, nerve cord, and muscles, and most of the brain, including eye and statocyst, are destroyed at the same time that the large pharyngeal cavity of the adult develops. A tadpole metamorphosing into an adult frog loses its tail—the cells of which are destroyed and devoured by phagocytic cells—its gills, and its larval mouthparts; concurrently the legs of the adult frog develop progressively, the structure of the mouth and alimentary canal change, and the skin acquires a bony (keratinized) layer and a system of subcutaneous glands.

The complicated changes taking place during metamorphosis, especially in the case of necrobiotic metamorphosis, must be performed in a coordinated way. So that no changes are made prematurely and no organ systems are left behind in the general transformation, some common signal for the change must be provided. For both insect and amphibian metamorphoses, which have been the most extensively studied, the signal is a hormonal one, sent in the blood to all the cells and tissues of the body.

Metamorphosis in an insect is complicated by the fact that the rigid cuticle covering its body is very restrictive; new features can appear only after a molt, when the old cuticle is replaced by a newly formed one. Molting in insects is caused by the action of two hormones. In the brain of insects, several groups of neurosecretory cells produce the first hormone. This brain hormone does not itself affect molting but stimulates the prothoracic gland, a loose mass of secretory cells situated in the thorax in close association with tracheal tubes. In response to the stimulation by the brain hormone, the prothoracic gland releases into the blood a second hormone, the molting hormone, or ecdysone. Under the influence of ecdysone, the tissues of the body produce a new cuticle under the old one, after which the old cuticle is shed (the actual molting). The new cuticle embodies any new developmental features that were scheduled to appear. The kind of feature that emerges after a molt is controlled by a third organ of internal secretion, the corpus allatum, secretory tissue situated posterior to the brain, near or around the dorsal aorta and usually appearing as a pair of separate or fused organs. The corpora allata emit the juvenile hormone, which, as long as it circulates in the blood, acts to perpetuate the larval form. As the larva approaches the end of its development, however, the corpora allata stop producing juvenile hormone or reduces its quantity; whereupon, the larva, at the next molt, metamorphoses into an adult. Withdrawal of the juvenile hormone is the immediate cause of metamorphosis, in conjunction with the brain hormone and ecdysone, which are responsible for the shedding of the larval cuticle and for the production of the new cuticle embodying the features of the imago. Metamorphosis through the stage of the pupa is affected by diminishing levels of juvenile hormone, which determine first the transformation of the larva into a pupa and, with further reduction of the juvenile-hormone level, the final step of transformation of the pupa into the adult.

The metamorphosis of a tadpole into a frog also depends upon two hormones: one initiating the process and the other directly influencing the tissues involved in the change. The first hormone is the thyrotropic hormone, produced by the hypophysis. It has no immediate effect on the tissues of the body but activates the thyroid gland to produce several substances, the most important of which is thyroxine. Thyroxine and other iodine-containing compounds circulate in the blood and cause changes that, in their entirety, constitute the process of metamorphosis. It is remarkable that different tissues react in different ways to the presence of thyroxine. The muscles of the tadpole's tail degenerate, whereas the muscles of the trunk and legs are not affected; in fact, the growth and development of limbs are stimulated as a part of metamorphosis. The effect of the hormone depends on the nature of the reacting cells and tissues—i.e., on their competence—just as the embryonic inductor in the earlier stages of development influences only cells with the competence for a particular kind of reaction.

Direct Development

If an animal after birth or emergence from an egg differs from the adult in comparatively minor details (apart from not having functional sex organs), the development is said to be direct. There is no larval stage and no metamorphosis. Direct development does not mean, however, that no changes occur between birth and adulthood. One very obvious change is the growth of the animal.

The rate of growth—not absolute increase—is highest in the early stages of postembryonic life; subsequently, growth continues to slow, ceasing completely at the attainment of adulthood. The rate of growth is dependent on many factors, both external (feeding, temperature) and internal. Of the internal factors, the most important are hormones, especially the growth hormone produced by the hypophysis. If the growth hormone is produced in insufficient quantities, the result is dwarfism; if it is produced in excessive quantities, the result is gigantism.

In the case of direct development, the most important change is the attainment of sexual maturity, which is achieved in several steps and involves the action of several hormones. The gonad rudiments and rudiments of the supporting parts of the reproductive system remain inactive long after birth. At the approach of adulthood, however, two sets of hormones come into action: hypophyseal hormones stimulate the gonads to activity, and gonadal hormones (produced by the gonads) cause the supporting sex organs and other sex characters to become fully developed. To become functional, the gonads must be acted upon by secretions from the hypophysis. In immature females the follicle-stimulating hormone, which alone causes the egg follicles and the oocytes to grow, and the luteinizing hormone stimulate the follicle cells to produce the female sex hormone, estrogen, which effects the development of the uterus, the milk glands, and other characteristics of the female sex. In the male, the same hypophyseal hormones are produced, with the result that the testes start to produce sperm and to secrete the male sex hormone, androgen. It appears that the luteinizing hormone is the more active in the male sex, being able to cause both spermatogenesis and androgen secretion. Androgen, in turn, brings about the development of the penis, the descent of the testicles before birth, the appearance of typical male hair growth, and other secondary sex characteristics.

Maturity and Death

Sexual maturity and the ensuing reproductive activity mark the pinnacle of development and morphogenesis and, for many animals, herald the end of life. The biological goal of the entire process is achieved with the launching of the next generation, and the life cycle that runs from the formation of gametes by one generation to the formation of gametes by the next generation is completed. In many animals the females die after laying their eggs; the males may have died earlier, after pairing. Indeed, some males (spiders, praying mantises) are eaten by the females immediately after copulation.

The developmental period can only truly be said to end with the termination of an organism, for much activity continues to unfold new developmental sequences, not all of them progressive and favourable, to be sure. Senescence, or a decline in abilities, signals advancing age in mammals but is not a general occurrence in the animal kingdom. Far more animals continue to function at near-peak capacity well into old age. And even among those species—salmons, eels, many moths—whose members die after a single reproductive act, death is relatively swift and not accompanied by a prolonged period of deterioration.

In most animals the reproductive potential is not exhausted in a single act of gamete production, but the sexually mature individuals remain alive and reproduce repeatedly. In these cases life may extend long beyond the first attainment of reproductive ability and be accompanied by further growth of the individuals, as occurs in most fishes, amphibians, and reptiles, and also in mollusks and certain other invertebrates. In the case of prolonged life spans, however, reproductive activity may cease with advancing age, and a senile involution take place, as is observed mainly in mammals and, particularly, in man. The changes taking place may be described as regressive development. In most animals, however, the end of life is not preceded by any overt traces of senility. As a general rule, then, the attainment of reproductive ability may be said to be the final phase of progressive development among animals. A gradual loss of alertness and vigour is typical of the aging pattern of primates and is especially important to man.

References

- Developmental-biology: microbenotes.com, Retrieved 09, August 2020
- Plant-development, science: britannica.com, Retrieved 02, May 2020
- 13-2-development-and-organogenesis, biology: opentextbc.ca, Retrieved 11, April 2020
- Biological-foundations-of-human-development, wmopen-lifespandevelopment: courses.lumenlearning.com, Retrieved 22, July 2020
- Animal-development, science: britannica.com, Retrieved 08, March 2020

Chapter 4

Significant Concepts in Evolutionary Biology

The concept of evolutionary biology when two or more species have a common ancestor as an effect of speciation is known as common descent. Several other concepts of evolutionary biology like speciation, modern evolutionary synthesis, adaptation and genetic variation are discussed in this chapter. It elucidates the varied processes associated with this area of study.

Common Descent

Descent describes how, in evolutionary biology, a group of organisms share a most recent common ancestor. There is evidence of common descent that all life on Earth is descended from the last universal ancestor. In July 2016, scientists reported identifying a set of 355 genes from the Last Universal Common Ancestor (LUCA) of all organisms living on Earth.

Common ancestry between organisms of different species arises during speciation, in which new species are established from a single ancestral population. Organisms which share a more recent common ancestor are more closely related. The most recent common ancestor of all currently living organisms is the last universal ancestor, which lived about 3.9 billion years ago. The two earliest evidences for life on Earth are graphite found to be biogenic in 3.7 billion-year-old metasedimentary rocks discovered in western Greenland and microbial mat fossils found in 3.48 billion-year-old sandstone discovered in Western Australia. All currently living organisms on Earth share a common genetic heritage (universal common descent), with each being the descendant from a single original species, though the suggestion of substantial horizontal gene transfer during early evolution has led to questions about monophyly of life.

Universal common descent through an evolutionary process was first proposed by the English naturalist Charles Darwin in *On the Origin of Species* (1859), which concluded: "There is grandeur in this view of life, with its several powers, having been originally breathed into a few forms or into one; and that, whilst this planet has gone cycling on according to the fixed law of gravity, from so simple a beginning endless forms most beautiful and most wonderful have been, and are being, evolved."

History

In the 1740s, French mathematician Pierre Louis Maupertuis made the first known suggestion in a series of essays that all organisms may have had a common ancestor, and that they had diverged through random variation and natural selection. In *Essai de cosmologie* (1750), Maupertuis noted:

May we not say that, in the fortuitous combination of the productions of Nature, since only those creatures *could* survive in whose organizations a certain degree of adaptation was present, there is nothing extraordinary in the fact that such adaptation is actually found in all these species which now exist? Chance, one might say, turned out a vast number of individuals; a small proportion of these were organized in such a manner that the animals' organs could satisfy their needs. A much greater number showed neither adaptation nor order; these last have all perished.... Thus the species which we see today are but a small part of all those that a blind destiny has produced.

In 1790, Immanuel Kant wrote in *Kritik der Urteilskraft* (*Critique of Judgement*) that the analogy of animal forms implies a common original type, and thus a common parent.

In 1794, Charles Darwin's grandfather, Erasmus Darwin, asked:

[W]ould it be too bold to imagine, that in the great length of time, since the earth began to exist, perhaps millions of ages before the commencement of the history of mankind, would it be too bold to imagine, that all warm-blooded animals have arisen from one living filament, which THE GREAT FIRST CAUSE endued with animality, with the power of acquiring new parts attended with new propensities, directed by irritations, sensations, volitions, and associations; and thus possessing the faculty of continuing to improve by its own inherent activity, and of delivering down those improvements by generation to its posterity, world without end?

Charles Darwin's views about common descent, as expressed in *On the Origin of Species*, were that it was possible that there was only one progenitor for all life forms:

Therefore I should infer from analogy that probably all the organic beings which have ever lived on this earth have descended from some one primordial form, into which life was first breathed.

Evidence of Universal Common Descent

Common Biochemistry and Genetic Code

All known forms of life are based on the same fundamental biochemical organization: genetic information encoded in DNA, transcribed into RNA, through the effect of protein- and RNA-enzymes, then translated into proteins by (highly similar) ribosomes, with ATP, NADPH and others as energy sources, etc. Furthermore, the genetic code (the "translation table" according to which DNA information is translated into proteins) is nearly identical for all known lifeforms, from bacteria and archaea to animals and plants. The universality of this code is generally regarded by biologists as definitive evidence in favor of the theory of universal common descent. Analysis of the small differences in the genetic code has also provided support for universal common descent. An example would be Cytochrome c which most organisms actually share. A statistical comparison of various alternative hypotheses has shown that universal common ancestry is significantly more probable than models involving multiple origins.

Selectively Neutral Similarities

Similarities which have no adaptive relevance cannot be explained by convergent evolution, and therefore they provide compelling support for the theory of universal common descent.

Such evidence has come from two areas: amino acid sequences and DNA sequences. Proteins with the same three-dimensional structure need not have identical amino acid sequences; any irrelevant similarity between the sequences is evidence for common descent. In certain cases, there are several codons (DNA triplets) that code for the same amino acid. Thus, if two species use the same codon at the same place to specify an amino acid that can be represented by more than one codon, that is evidence for a recent common ancestor.

Other Similarities

The universality of many aspects of cellular life is often pointed to as supportive evidence to the more compelling evidence listed above. These similarities include the energy carrier adenosine triphosphate (ATP), and the fact that all amino acids found in proteins are left-handed. It is, however, possible that these similarities resulted because of the laws of physics and chemistry, rather than universal common descent and therefore resulted in convergent evolution.

Phylogenetic Trees

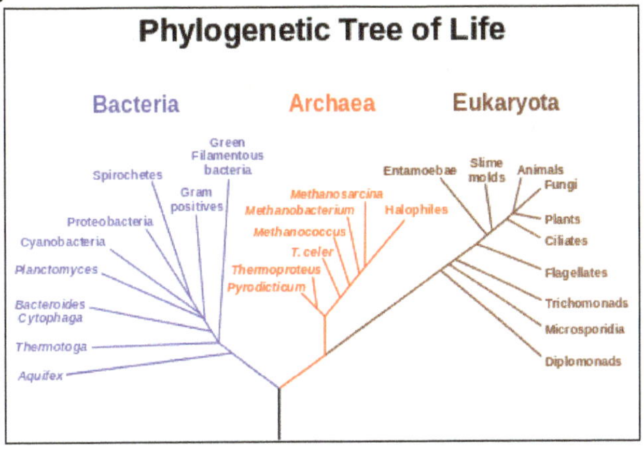

A phylogenetic tree based on rRNA genes.

Another important piece of evidence is that it is possible to construct detailed phylogenetic trees (i.e., "genealogic trees" of species) mapping out the proposed divisions and common ancestors of all living species. In 2010, Douglas L. Theobald published a statistical analysis of available genetic data, mapping them to phylogenetic trees, that gave "strong quantitative support, by a formal test, for the unity of life." It should be noted, however, that the "formal test" was criticised for not including consideration of convergent evolution, and Theobald has defended the method against this claim.

Traditionally, these trees have been built using morphological methods, such as appearance, embryology, etc. Recently, it has been possible to construct these trees using molecular data, based on similarities and differences between genetic and protein sequences. All these methods produce essentially similar results, even though most genetic variation has no influence over external morphology. That phylogenetic trees based on different types of information agree with each other is strong evidence of a real underlying common descent.

Illustrations of Common Descent

Artificial Selection

Artificial selection demonstrates the diversity that can exist among organisms that share a relatively recent common ancestor. In artificial selection, humans selectively direct the breeding of one species at each generation, allowing only those organisms that exhibit desired characteristics to reproduce. These characteristics become increasingly well-developed in successive generations. Artificial selection was successful long before science discovered the genetic basis.

Dog Breeding

The Chihuahua mix and Great Dane both share a common ancestor, the wolf, but show the power of artificial selection to create diversity of form in a relatively short period of time.

The diversity of domesticated dogs is an example of the power of artificial selection. All breeds share common ancestry, having descended from wolves. Humans selectively bred them to enhance specific characteristics, such as color and length or body size. This created a range of breeds that include the Chihuahua, Great Dane, Basset Hound, Pug, and Poodle. Wild wolves, which did not undergo artificial selection, are relatively uniform in comparison.

Wild cabbage

Early farmers cultivated many popular vegetables from the *Brassica oleracea* (wild cabbage) by artificially selecting for certain attributes. Common vegetables such as cabbage, kale, broccoli, cauliflower, kohlrabi and Brussels sprouts are all descendants of the wild cabbage plant. Brussels sprouts were created by artificially selecting for large bud size. Broccoli was bred by selecting for large flower stalks. Cabbage was created by selecting for short petioles. Kale was bred by selecting for large leaves.

Natural Selection

Natural selection is the evolutionary process by which heritable traits that increase an individual's fitness become more common, and heritable traits that decrease an individual's fitness become less common.

Darwin's Finches

During his studies on the Galápagos Islands, Charles Darwin observed 13 species of finches that are closely related and differ most markedly in the shape of their beaks. The beak of each species is suited to the food available in its particular environment, suggesting that beak shapes evolved by natural selection. Large beaks were found on the islands where the primary source of food for the finches are nuts and therefore the large beaks allowed the birds to be better equipped for opening the nuts and staying well nourished. Slender beaks were found on the finches which found insects to be the best source of food on the island they inhabited; their slender beaks allowed the birds to be better equipped for pulling out the insects from their tiny hiding places. The finch is also found on the main land and it is thought that they migrated to the islands and began adapting to their environment through natural selection.

Speciation

Speciation is the evolutionary process by which reproductively isolated biological populations evolve to become distinct species. The biologist Orator F. Cook was the first to coin the term 'speciation' for the splitting of lineages or "cladogenesis," as opposed to "anagenesis" or "phyletic evolution" occurring within lineages. Charles Darwin was the first to describe the role of natural selection in speciation.

There are four geographic modes of speciation in nature, based on the extent to which speciating populations are isolated from one another: allopatric, peripatric, parapatric, and sympatric. Speciation may also be induced artificially, through animal husbandry, agriculture, or laboratory experiments. Whether genetic drift is a minor or major contributor to speciation is the subject matter of much ongoing discussion.

It is widely appreciated that sexual selection could drive speciation in many clades, independently of natural selection. However the term "speciation", in this context, tends to be used in two different, but not mutually exclusive senses. The first and most commonly used sense refers to the "birth" of new species. That is, the splitting of an existing species into two separate species, or the budding off of a new species from a parent species, both driven by a biological "fashion fad" (a preference for a feature, or features, in one or both sexes, that do not necessarily have any adaptive qualities). In the second sense, "speciation" refers the wide-spread tendency of sexual creatures to be grouped into clearly defined species, rather than forming a continuum of phenotypes both in time and space - which would be the more obvious or logical consequence of natural selection. This was indeed recognized by Darwin as problematic, and included in his *On the Origin of Species* (1859), under the heading "Difficulties with the Theory". There are several suggestions as to how mate choice might play a significant role in resolving Darwin's dilemma.

Modes of Speciation

All forms of natural speciation have taken place over the course of evolution; however, debate persists as to the relative importance of each mechanism in driving biodiversity.

One example of natural speciation is the diversity of the three-spined stickleback, a marine fish that, after the last glacial period, has undergone speciation into new freshwater colonies in isolated lakes and streams. Over an estimated 10,000 generations, the sticklebacks show structural differences that are greater than those seen between different genera of fish including variations in fins, changes in the number or size of their bony plates, variable jaw structure, and color differences.

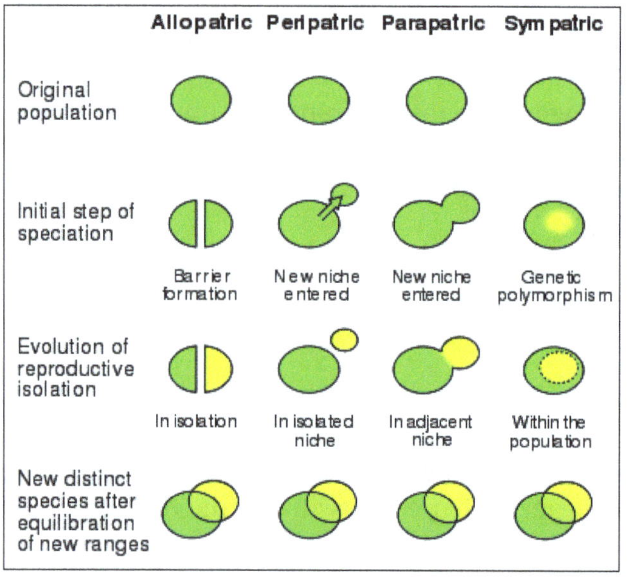

Comparison of allopatric, peripatric, parapatric and sympatric speciation.

Allopatric

During allopatric (from the ancient Greek *allos*, "other" + Greek *patrā*, "fatherland") speciation, a population splits into two geographically isolated populations (for example, by habitat fragmentation due to geographical change such as mountain formation). The isolated populations then undergo genotypic and/or phenotypic divergence as: (a) they become subjected to dissimilar selective pressures; (b) they independently undergo genetic drift; (c) different mutations arise in the two populations. When the populations come back into contact, they have evolved such that they are reproductively isolated and are no longer capable of exchanging genes. Island genetics is the term associated with the tendency of small, isolated genetic pools to produce unusual traits. Examples include insular dwarfism and the radical changes among certain famous island chains, for example on Komodo. The Galápagos Islands are particularly famous for their influence on Charles Darwin. During his five weeks there he heard that Galápagos tortoises could be identified by island, and noticed that finches differed from one island to another, but it was only nine months later that he reflected that such facts could show that species were changeable. When he returned to England, his speculation on evolution deepened after experts informed him that these were separate species, not just varieties, and famously that other differing Galápagos birds were all species of finches. Though the finches were less important for Darwin, more recent research has shown the birds now known as Darwin's finches to be a classic case of adaptive evolutionary radiation.

Peripatric

In peripatric speciation, a subform of allopatric speciation, new species are formed in isolated, smaller peripheral populations that are prevented from exchanging genes with the main population. It is related to the concept of a founder effect, since small populations often undergo bottlenecks. Genetic drift is often proposed to play a significant role in peripatric speciation.

Case Studies:

- Mayr bird fauna
- The Australian bird *Petroica multicolor*
- Reproductive isolation occurs in populations of *Drosophila* subject to population bottlenecking

Parapatric

In parapatric speciation, there is only partial separation of the zones of two diverging populations afforded by geography; individuals of each species may come in contact or cross habitats from time to time, but reduced fitness of the heterozygote leads to selection for behaviours or mechanisms that prevent their interbreeding. Parapatric speciation is modelled on continuous variation within a "single," connected habitat acting as a source of natural selection rather than the effects of isolation of habitats produced in peripatric and allopatric speciation.

Parapatric speciation may be associated with differential landscape-dependent selection. Even if there is a gene flow between two populations, strong differential selection may impede assimilation and different species may eventually develop. Habitat differences may be more important in the development of reproductive isolation than the isolation time. Caucasian rock lizards *Darevskia rudis*, *D. valentini* and *D. portschinskii* all hybridize with each other in their hybrid zone; however, hybridization is stronger between *D. portschinskii* and *D. rudis*, which separated earlier but live in similar habitats than between *D. valentini* and two other species, which separated later but live in climatically different habitats.

Ecologists refer to parapatric and peripatric speciation in terms of ecological niches. A niche must be available in order for a new species to be successful.

Case Studies:

- Ring species
    - The *Larus* gulls form a ring species around the North Pole.
    - The *Ensatina* salamanders, which form a ring round the Central Valley in California.
    - The greenish warbler (*Phylloscopus trochiloides*), around the Himalayas.
- The grass *Anthoxanthum* has been known to undergo parapatric speciation in such cases as mine contamination of an area.

Sympatric

Sympatric speciation refers to the formation of two or more descendant species from a single ancestral species all occupying the same geographic location.

Japanese rice fish.

Freshwater angelfish, a cichlid.

Often-cited examples of sympatric speciation are found in insects that become dependent on different host plants in the same area. However, the existence of sympatric speciation as a mechanism of speciation remains highly debated.

The best illustrated example of sympatric speciation is that of the cichlids of East Africa inhabiting the Rift Valley lakes, particularly Lake Victoria, Lake Malawi and Lake Tanganyika. There are over 800 described species, and according to estimate, there could be well over 1,600 species in the region. All the species have diversified from a common ancestral fish, the Japanese rice fish (*Oryzias latipes*) about 113 million years ago. Their evolution is cited as an example of both natural and sexual selection. A 2008 study suggests that sympatric speciation has occurred in Tennessee cave salamanders. Sympatric speciation driven by ecological factors may also account for the extraordinary diversity of crustaceans living in the depths of Siberia's Lake Baikal.

Budding speciation has been proposed as a particular form of sympatric speciation, whereby small groups of individuals become progressively more isolated from the ancestral stock by breeding preferentially with one another. This type of speciation would be driven by the conjunction of various advantages of inbreeding such as the expression of advantageous recessive phenotypes, reducing the recombination load, and reducing the cost of sex.

Example: Hawthorn fly

One example of evolution at work is the case of the hawthorn fly (*Rhagoletis pomonella*), also known as the apple maggot fly, which appears to be undergoing sympatric speciation. Different populations of hawthorn fly feed on different fruits. A distinct population emerged in North America in the 19th century some time after apples, a non-native species, were introduced. This apple-feeding population normally feeds only on apples and not on the historically preferred fruit of hawthorns. The current hawthorn feeding population does not normally feed on apples. Some evidence, such as the fact that six out of thirteen allozyme loci are different, that hawthorn flies mature later in the season and take longer to mature than apple flies; and that there is little evidence of interbreeding (researchers have documented a 4-6% hybridization rate) suggests that sympatric speciation is occurring. The emergence of the new hawthorn fly is an example of evolution in progress.

Rhagoletis pomonella.

Example: Three-spined Sticklebacks

The three-spined stickleback (*Gasterosteus aculeatus*).

Freshwater three-spined sticklebacks, which have been studied by Dolph Schluter, were once thought to provide an intriguing example best explained by sympatric speciation. Schluter and colleagues found two different species of three-spined sticklebacks in each of five different lakes. Each lake contained a large benthic species with a large mouth that feeds on large prey in the littoral zone, as well as a smaller limnetic species with a smaller mouth that feeds on the small plankton in open water. DNA analysis indicated that each lake was colonized independently, presumably by a marine ancestor, after the last glacial period. It also showed that the two species in each lake are more closely related to each other than they are to any of the species in the other lakes.The two species in each lake are reproductively isolated; neither mates with the other. However, aquarium tests showed that the benthic species from one lake is able to mate with the benthic species from the other lakes. Likewise the limnetic species from the different lakes are able to mate with each other. These benthic and limnetic species even display their mating preferences when presented with sticklebacks from Japanese lakes. A Canadian benthic prefers a Japanese benthic over its close limnetic relative from its own lake.

The researchers concluded that in each lake there had been great competition within a single original species for limited resources. This led to disruptive selection—competition favoring fishes at either extreme of body size and mouth size over those nearer the mean, as well as assortative mating—each size preferred mates like it. The result was a divergence into two subpopulations exploiting different food in different parts of the lake.The fact that this pattern of speciation occurred the same way on three separate occasions suggests strongly that ecological factors in a sympatric population can cause speciation.

However, the DNA evidence cited above is from mitochondrial DNA (mtDNA), which can often move easily between closely related species ("introgression") when they hybridize or engage in despeciation. A more recent study, using genetic markers from the nuclear genome, shows that limnetic forms in different lakes are more closely related to each other (and to marine lineages) than to benthic forms in the same lake. The three-spine stickleback is now usually considered an example of "double invasion" (a form of allopatric speciation) in which repeated invasions of marine forms have subsequently differentiated into benthic and limnetic forms. The three-spine stickleback provides an example of how molecular biogeographic studies that rely solely on mtDNA can be misleading, and that consideration of the genealogical history of alleles from multiple unlinked markers (i.e., nuclear genes) is necessary to infer speciation histories.

Reinforcement

Reinforcement, also called the *Wallace effect*, is the process by which natural selection increases reproductive isolation. It may occur after two populations of the same species are separated and then come back into contact. If their reproductive isolation was complete, then they will have already developed into two separate incompatible species. If their reproductive isolation is incomplete, then further mating between the populations will produce hybrids, which may or may not be fertile. If the hybrids are infertile, or fertile but less fit than their ancestors, then there will be further reproductive isolation and speciation has essentially occurred (e.g., as in horses and donkeys.)

The reasoning behind this is that if the parents of the hybrid offspring each have naturally selected traits for their own certain environments, the hybrid offspring will bear traits from both, therefore would not fit either ecological niche as well as either parent. The low fitness of the hybrids would cause selection to favor assortative mating, which would control hybridization. This is sometimes called the Wallace effect after the evolutionary biologist Alfred Russel Wallace who suggested in the late 19th century that it might be an important factor in speciation. Conversely, if the hybrid offspring are more fit than their ancestors, then the populations will merge back into the same species within the area they are in contact.

Reinforcement favoring reproductive isolation is required for both parapatric and sympatric speciation. Without reinforcement, the geographic area of contact between different forms of the same species, called their "hybrid zone," will not develop into a boundary between the different species. Hybrid zones are regions where diverged populations meet and interbreed. Hybrid offspring are very common in these regions, which are usually created by diverged species coming into secondary contact. Without reinforcement, the two species would have uncontrollable inbreeding. Reinforcement may be induced in artificial selection experiments as described below.

Ecological and Parallel Speciation

Ecological selection is "the interaction of individuals with their environment during resource acquisition". Natural selection is inherently involved in the process of speciation, whereby, "under ecological speciation, populations in different environments, or populations exploiting different resources, experience contrasting natural selection pressures on the traits that directly or indirectly bring about the evolution of reproductive isolation". Evidence for the role ecology plays in the process of speciation exists. Studies of stickleback populations support ecologically-linked speciation arising as a by-product, alongside numerous studies of parallel speciation.

Parallel speciation is where "greater reproductive isolation repeatedly evolves between independent populations adapting to contrasting environments than between independent populations adapting to similar environments". It is established that ecological speciation occurs and with much of the evidence, "...accumulated from top-down studies of adaptation and reproductive isolation".

Artificial Speciation

European mouflon (*Ovis aries musimon*).

New species have been created by domesticated animal husbandry, but the initial dates and methods of the initiation of such species are not clear. Often, the domestic counterpart of the wild ancestor can still interbreed and produce fertile offspring as in the case of domestic cattle, that can be considered the same species as several varieties of wild ox, gaur, yak, etc., or domestic sheep that can interbreed with the mouflon.

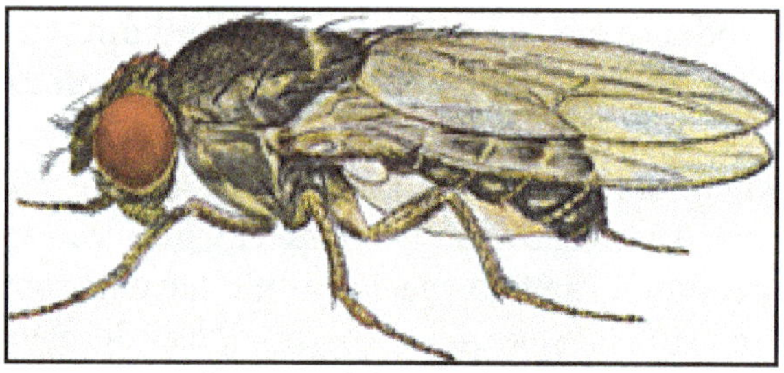

Male *Drosophila pseudoobscura*.

The best-documented creations of new species in the laboratory were performed in the late 1980s. William R. Rice and George W. Salt bred *Drosophila melanogaster* fruit flies using a maze with three different choices of habitat such as light/dark and wet/dry. Each generation was placed into the maze, and the groups of flies that came out of two of the eight exits were set apart to breed with each other in their respective groups. After thirty-five generations, the two groups and their offspring were isolated reproductively because of their strong habitat preferences: they mated only within the areas they preferred, and so did not mate with flies that preferred the other areas. The history of such attempts is described by Rice and Elen E. Hostert (1993).

Diane Dodd used a laboratory experiment to show how reproductive isolation can evolve in *Drosophila pseudoobscura* fruit flies after several generations by placing them in different media, starch- and maltose-based media.

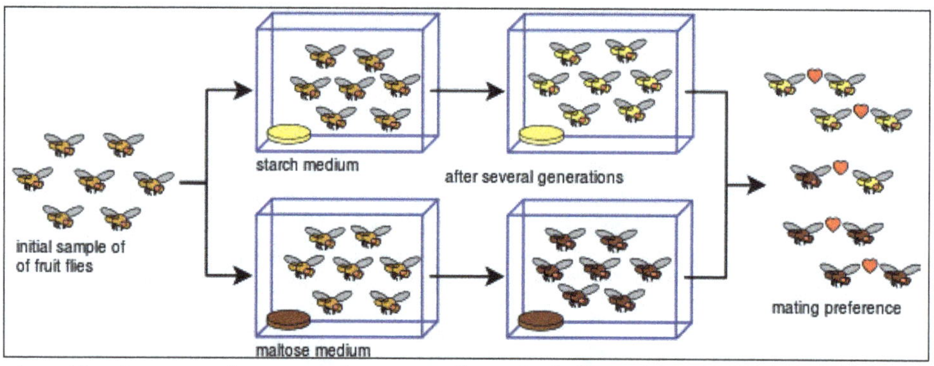

Dodd's experiment has been easy for many others to replicate, including with other kinds of fruit flies and foods. Research in 2005 has shown that this rapid evolution of reproductive isolation may in fact be a relic of infection by *Wolbachia* bacteria.

Alternatively, these observations are consistent with the notion that sexual creatures are inherently reluctant to mate with individuals whose appearance or behavior is different from the norm. The risk that such deviations are due to heritable maladaptations is very high. Thus, if a sexual creature, unable to predict natural selection's future direction, is conditioned to produce the fittest offspring possible, it will avoid mates with unusual habits or features. Sexual creatures will then inevitably tend to group themselves into reproductively isolated species.

Genetics

Few speciation genes have been found. They usually involve the reinforcement process of late stages of speciation. In 2008, a speciation gene causing reproductive isolation was reported. It causes hybrid sterility between related subspecies. The order of speciation of three groups from a common ancestor may be unclear or unknown; a collection of three such species is referred to as a "trichotomy."

Speciation via Polyploidization

Polyploidy is a mechanism that has caused many rapid speciation events in sympatry because offspring of, for example, tetraploid x diploid matings often result in triploid sterile progeny. However, not all polyploids are reproductively isolated from their parental plants, and gene flow may still occur for example through triploid hybrid x diploid matings that produce tetraploids, or matings between meiotically unreduced gametes from diploids and gametes from tetraploids.

It has been suggested that many of the existing plant and most animal species have undergone an event of polyploidization in their evolutionary history. Reproduction of successful polyploid species is sometimes asexual, by parthenogenesis or apomixis, as for unknown reasons many asexual organisms are polyploid. Rare instances of polyploid mammals are known, but most often result in prenatal death.

Hybrid Speciation

Hybridization between two different species sometimes leads to a distinct phenotype. This phenotype can also be fitter than the parental lineage and as such natural selection may then favor these individuals. Eventually, if reproductive isolation is achieved, it may lead to a separate species. However, reproductive isolation between hybrids and their parents is particularly difficult to achieve and thus hybrid speciation is considered an extremely rare event. The mariana mallard is thought to have arisen from hybrid speciation.

Hybridization is an important means of speciation in plants, since polyploidy (having more than two copies of each chromosome) is tolerated in plants more readily than in animals. Polyploidy is important in hybrids as it allows reproduction, with the two different sets of chromosomes each being able to pair with an identical partner during meiosis. Polyploids also have more genetic diversity, which allows them to avoid inbreeding depression in small populations.

Hybridization without change in chromosome number is called homoploid hybrid speciation. It is considered very rare but has been shown in *Heliconius* butterflies and sunflowers. Polyploid speciation, which involves changes in chromosome number, is a more common phenomenon, especially in plant species.

Gene Transposition

Theodosius Dobzhansky, who studied fruit flies in the early days of genetic research in 1930s, speculated that parts of chromosomes that switch from one location to another might cause a species to split into two different species. He mapped out how it might be possible for sections of chromosomes to relocate themselves in a genome. Those mobile sections can cause sterility in inter-species hybrids, which can act as a speciation pressure. In theory, his idea was sound, but scientists long debated whether it actually happened in nature. Eventually a competing theory involving the gradual accumulation of mutations was shown to occur in nature so often that geneticists largely dismissed the moving gene hypothesis.

However, 2006 research shows that jumping of a gene from one chromosome to another can contribute to the birth of new species. This validates the reproductive isolation mechanism, a key component of speciation.

Historical Background

Life Timeline

In addressing the question of the origin of species, there are two key issues: (1) what are the evolutionary mechanisms of speciation, and (2) what accounts for the separateness and individuality of species in the biota? Since Charles Darwin's time, efforts to understand the nature of species have primarily focused on the first aspect, and it is now widely agreed that the critical factor behind the origin of new species is reproductive isolation. Next we focus on the second aspect of the origin of species.

Darwin's Dilemma: Why Do Species Exist?

In *On the Origin of Species* (1859), Darwin interpreted biological evolution in terms of natural selection, but was perplexed by the clustering of organisms into species. Chapter 6 of Darwin's book is entitled "Difficulties of the Theory." In discussing these "difficulties" he noted "Firstly, why, if species have descended from other species by insensibly fine gradations, do we not everywhere see innumerable transitional forms? Why is not all nature in confusion instead of the species being, as we see them, well defined?" This dilemma can be referred to as the absence or rarity of transitional varieties in habitat space.

Another dilemma, related to the first one, is the absence or rarity of transitional varieties in time. Darwin pointed out that by the theory of natural selection "innumerable transitional forms must have existed," and wondered "why do we not find them embedded in countless numbers in the crust of the earth." That clearly defined species actually do exist in nature in both space and time implies that some fundamental feature of natural selection operates to generate and maintain species.

The Effect of Sexual Reproduction on Species Formation

It has been argued that the resolution of Darwin's dilemmas lies in the fact that out-crossing sexual reproduction has an intrinsic cost of rarity. The cost of rarity arises as follows. If, on a resource gradient, a large number of separate species evolve, each exquisitely adapted to a very narrow band on that gradient, each species will, of necessity, consist of very few members. Finding a mate under these circumstances may present difficulties when many of the individuals in the neighborhood belong to other species. Under these circumstances, any species that happens, by chance, to increase in population size (at the expense of one or other of its neighboring species, if the environment is saturated), this will immediately make it easier for its members to find sexual partners. The members of the neighboring species, whose population sizes have decreased, will experience greater difficulty in finding mates, and therefore form pairs less frequently than in the larger species. This has a snowball effect, with large species growing at the expense of the smaller, rarer species, eventually driving them to extinction. Eventually, only a few species remain, each distinctly different from the other. The cost of rarity not only involves the costs of failure to find a mate, but also indirect costs such as the cost of communication in seeking out a partner at low population densities.

A photograph of an African pygmy kingfisher, showing details of appearance and coloration that are shared by adults of that species to a high degree of fidelity.

Bernstein *et al.* argue furthermore that if an environmental gradient is populated by a single species which is perfectly adapted to only a small portion of that environment, it will be difficult for better-adapted individuals to pass their adaptation on to others in region. Such advantageous characteristics are unlikely to be due to a single altered gene, but rather to a combination of several altered genes, each of which, on its own, imparts little or no benefit to its carrier. If such an individual mates with a randomly selected mate, the advantageous combination of genes will be broken up, and the advantage lost, unless it happens to mate with another individual with the same advantageous combination of altered genes. This will be an exceptionally rare event, the consequence of which is that the species will be resistant to change over time or to the budding off of new species.

Rarity brings with it other costs. A rare or unusual feature is very seldom advantageous. In most instances, it will be indicative of a (non-silent) mutation, which is almost certain to be deleterious. It therefore behooves sexual creatures to avoid mates sporting rare or unusual features. Should this be the case, then sexual populations will rapidly shed rare or peripheral phenotypic features, thus canalizing the entire external appearance, as illustrated in the accompanying illustration of the African pygmy kingfisher, *Ispidina picta*. This remarkable uniformity of all the adult members of a sexual species has stimulated the proliferation of field guides on birds, mammals, reptiles, insects, and many other taxons, in which each species can be described in full by means of a single illustration (or a pair of illustrations if there is sufficient sexual dimorphism). Once a population has become as homogeneous in appearance as is typical of most species (and is illustrated in the photograph of the African pygmy kingfisher), its members will avoid mating with members of other populations that look different from themselves. Thus, the avoidance of mates displaying rare and unusual phenotypic features inevitably leads to reproductive isolation, one of the hallmarks of speciation.

In the contrasting case of organisms that reproduce asexually, there is no cost of rarity; consequently, there are only benefits to fine-scale adaptation. Thus, asexual organisms very frequently show the continuous variation in form (often in many different directions) that Darwin expected evolution to produce, making their classification into "species" (more correctly, "morphospecies") very difficult.

Rates of Speciation

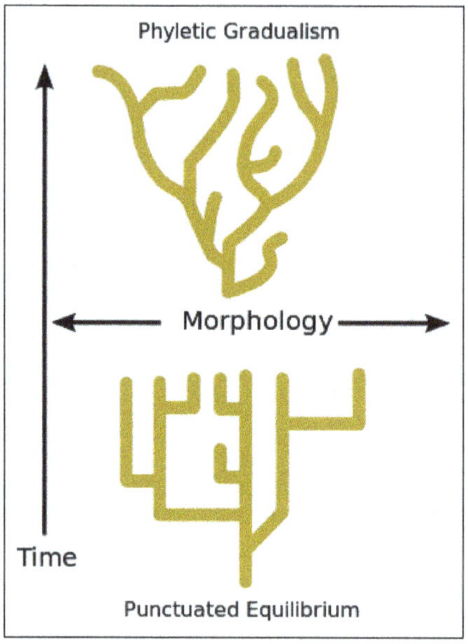

Phyletic gradualism, above, consists of relatively slow change over geological time. Punctuated equilibrium, bottom, consists of morphological stability and rare, relatively rapid bursts of evolutionary change.

There is debate as to the rate at which speciation events occur over geologic time. While some evolutionary biologists claim that speciation events have remained relatively constant and gradual over time, some palaeontologists such as Niles Eldredge and Stephen Jay Gould have argued that species usually remain unchanged over long stretches of time, and that speciation occurs only over relatively brief intervals, a view known as *punctuated equilibrium*.)

Punctuated Evolution

Evolution can be extremely rapid, as shown in the creation of domesticated animals and plants in a very short geological space of time, spanning only a few tens of thousands of years. Maize (*Zea mays*), for instance, was created in Mexico in only a few thousand years, starting about 7,000 to 12,000 years ago. This raises the question of why the long term rate of evolution is far slower than is theoretically possible.

Plants and Domestic Animals can Differ Markedly from their Wild Ancestors

Top: wild teosinte; middle: maize-teosinte hybrid; bottom: maize.

Ancestral wild cabbage.	Domesticated cauliflowe.

Ancestral Prussian carp.	Domestic goldfish.

Ancestral mouflon.	Domestic sheep.

Evolution is imposed on species or groups. It is not planned or striven for in some Lamarckist way. The mutations on which the process depends are random events, and, except for the "silent mutations" which do not affect the functionality or appearance of the carrier, are thus usually disadvantageous, and their chance of proving to be useful in the future is vanishingly small. Therefore, while a species or group might benefit from being able to adapt to a new environment by accumulating a wide range of genetic variation, this is to the detriment of the *individuals* who have to carry these mutations until a small, unpredictable minority of them ultimately contributes to such an adaptation. Thus, the *capability* to evolve is a group adaptation, a concept discredited by (for example) George C. Williams, John Maynard Smith and Richard Dawkins as selectively disadvantageous to the individual.

If sexual creatures avoid mutant mates with strange or unusual characteristics, then mutations that affect the external appearance of their carriers would seldom be passed on to the next and subsequent generations. They would therefore seldom be tested by natural selection. Evolution is, therefore, effectively halted or slowed down considerably. The only mutations that can accumulate in a population, on this punctuated equilibrium view, are ones that have no noticeable effect on the outward appearance and functionality of their bearers (i.e., they are "silent" or "neutral mutations," which can be, and are, used to trace the relatedness and age of populations and species.)

This argument implies that evolution can only occur if mutant mates cannot be avoided, as a result of a severe scarcity of potential mates. This is most likely to occur in small, isolated communities. These occur most commonly on small islands, in remote valleys, lakes, river systems, or caves, or during the aftermath of a mass extinction. Under these circumstances, not only is the choice of mates severely restricted but population bottlenecks, founder effects, genetic drift and inbreeding cause rapid, random changes in the isolated population's genetic composition. Furthermore, hybridization with a related species trapped in the same isolate might introduce additional genetic changes. If an isolated population such as this survives its genetic upheavals, and subsequently expands into an unoccupied niche, or into a niche in which it has an advantage over its competitors, a new species, or subspecies, will have come in being. In geological terms this will be an abrupt event. A resumption of avoiding mutant mates will thereafter result, once again, in evolutionary stagnation.

In apparent confirmation of this punctuated equilibrium view of evolution, the fossil record of an evolutionary progression typically consists of species that suddenly appear, and ultimately disappear, in many cases close to a million years later, without any change in external appearance. Graphically, these fossil species are represented by horizontal lines, whose lengths depict how long each of them existed. The horizontality of the lines illustrates the unchanging appearance of each of the fossil species depicted on the graph. During each species' existence new species appear at random intervals, each also lasting many hundreds of thousands of years before disappearing without a change in appearance. The exact relatedness of these concurrent species is generally impossible to determine. This is illustrated in the following diagram depicting the evolution of modern humans from the time that the hominins separated from the line that led to the evolution of our closest living primate relatives, the chimpanzees.

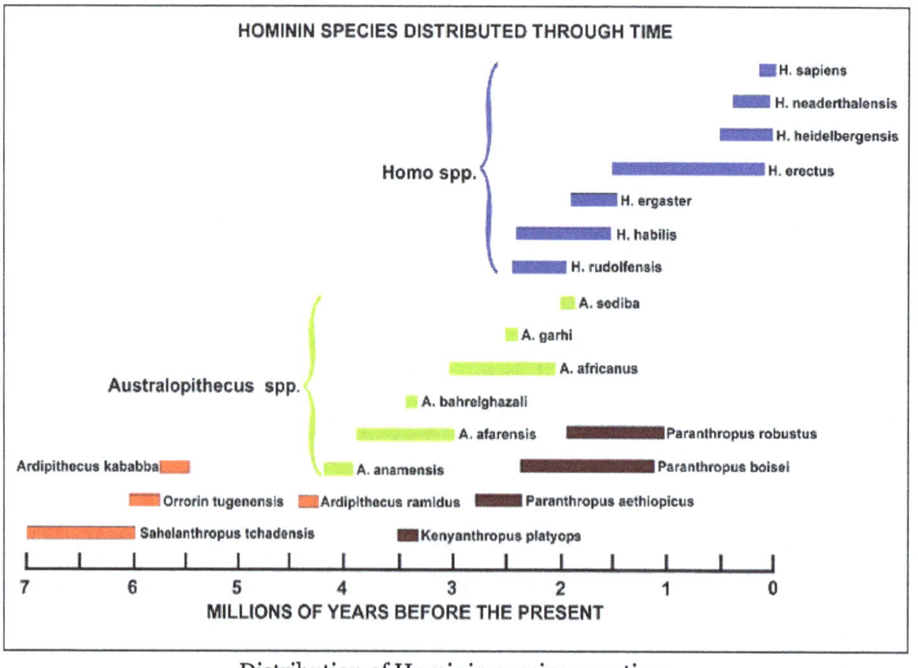

Distribution of Hominin species over time.

For similar evolutionary time lines see, for instance, the paleontological list of African dinosaurs, Asian dinosaurs, the Lampriformes and Amiiformes.

Modern Evolutionary Synthesis

The modern evolutionary synthesis (known as the new synthesis, the modern synthesis, the evolutionary synthesis, millennium synthesis or the neo-Darwinian synthesis) is a 20th-century synthesis of ideas from several fields of biology that provides an account of evolution which is widely accepted as the current paradigm in evolutionary biology, and reflects the consensus about how evolution works.

J. B .S. Haldane.

The 19th Century ideas of natural selection by Charles Darwin and Mendelian genetics by Gregor Mendel were united by Ronald Fisher, one of the three founders of population genetics, along with J. B. S. Haldane and Sewall Wright, between 1918 and 1932.

Ernst Mayr.

The modern synthesis solved difficulties and confusions caused by the specialisation and poor communication between biologists in the early years of the 20th century. At its heart was the question of whether Mendelian genetics could be reconciled with gradual evolution by means of natural selection. A second issue was whether the broad-scale changes of macroevolution seen by palaeontologists could be explained by changes seen in the microevolution of local populations .

The synthesis included evidence from biologists, trained in genetics, who studied populations in the field and in the laboratory. These studies were crucial to evolutionary theory. The synthesis drew together ideas from several branches of biology which had become separated, particularly genetics, cytology, systematics, botany, morphology, ecology and paleontology.

Julian Huxley invented the term in his 1942 book, *Evolution: The Modern Synthesis*. Major figures in the modern synthesis include, Theodosius Dobzhansky, E. B. Ford, Ernst Mayr, Bernhard Rensch, Sergei Chetverikov, George Gaylord Simpson, and G. Ledyard Stebbins.

Summary of the Modern Synthesis

The modern synthesis bridged the gap between the work of experimental geneticists and naturalists, and paleontologists. It states that:

1. All evolutionary phenomena can be explained in a way consistent with known genetic mechanisms and the observational evidence of naturalists.
2. Evolution is gradual: small genetic changes regulated by natural selection accumulate over long periods. Discontinuities amongst species (or other taxa) are explained as originating gradually through geographical separation and extinction. This theory contrasts with the saltation theory of William Bateson (1894).
3. Natural selection is by far the main mechanism of change; even slight advantages are important when continued. The object of selection is the phenotype in its surrounding environment.
4. The role of genetic drift is equivocal. Though strongly supported initially by Dobzhansky, it was downgraded later as results from ecological genetics were obtained.
5. Thinking in terms of populations, rather than individuals, is primary: the genetic diversity existing in natural populations is a key factor in evolution. The strength of natural selection in the wild is greater than previously expected; the effect of ecological factors such as niche occupation and the significance of barriers to gene flow are all important.
6. In palaeontology, the ability to explain historical observations by extrapolation from microevolution to macroevolution is proposed. Historical contingency means explanations at different levels may exist. Gradualism does not mean constant rate of change.

The idea that speciation occurs after populations are reproductively isolated has been much debated. In plants, polyploidy must be included in any view of speciation. Formulations such as 'evolution consists primarily of changes in the frequencies of alleles between one generation and another' were proposed rather later. The traditional view is that evolutionary developmental biology (evo-devo) played little part in the synthesis, but an account of Gavin de Beer's work by Stephen J. Gould suggests he may be an exception.

Developments Leading up to the Synthesis

1859–1899

Charles Darwin's On the Origin of Species (1859) was successful in convincing most biologists that evolution had occurred, but was less successful in convincing them that natural selection was its primary mechanism. In the 19th and early 20th centuries, variations of Lamarckism, orthogenesis ('progressive' evolution), and saltationism (evolution by jumps) were discussed as alternatives. Also, Darwin did not offer a precise explanation of how new species arise. As part of the disagreement about whether natural selection alone was sufficient to explain speciation, George Romanes coined the term neo-Darwinism to refer to the version of evolution advocated by Alfred Russel Wallace and August Weismann with its heavy dependence on natural selection. Weismann and Wallace rejected the Lamarckian idea of inheritance of acquired characteristics, something that Darwin had not ruled out.

Weismann's idea was that the relationship between the hereditary material, which he called the germ plasm (German, Keimplasma), and the rest of the body (the soma) was a one-way relationship: the germ-plasm formed the body, but the body did not influence the germ-plasm, except indirectly in its participation in a population subject to natural selection. Weismann was translated into English, and though he was influential, it took many years for the full significance of his work to be appreciated. Later, after the completion of the modern synthesis, the term neo-Darwinism came to be associated with its core concept: evolution, driven by natural selection acting on variation produced by genetic mutation, and genetic recombination (chromosomal crossovers).

1900–1915

Gregor Mendel's work was re-discovered by Hugo de Vries and Carl Correns in 1900. News of this reached William Bateson in England, who reported on the paper during a presentation to the Royal Horticultural Society in May 1900. It showed that the contributions of each parent retained their integrity rather than blending with the contribution of the other parent. This reinforced a division of thought, which was already present in the 1890s. The two schools were:

- Saltationism (large mutations or jumps), favored by early Mendelians who viewed hard inheritance as incompatible with natural selection
- Biometric school: led by Karl Pearson and Walter Weldon, argued vigorously against it, saying that empirical evidence indicated that variation was continuous in most organisms, not discrete as Mendelism predicted

The relevance of Mendelism to evolution was unclear and hotly debated, especially by Bateson, who opposed the biometric ideas of his former teacher Weldon. Many scientists believed the two theories substantially contradicted each other. This debate between the biometricians and the Mendelians continued for some 20 years and was only solved by the development of population genetics.

Thomas Hunt Morgan began his career in genetics as a saltationist, and started out trying to demonstrate that mutations could produce new species in fruit flies. However, the experimental work at his lab with the common fruit fly, *Drosophila melanogaster*, which helped establish the link between Mendelian genetics and the chromosomal theory of inheritance, demonstrated that rather than creating new species in a single step, mutations increased the genetic variation in the population.

The Foundation of Population Genetics

The first step towards the synthesis was the development of population genetics. R. A. Fisher, J. B. S. Haldane, and Sewall Wright provided critical contributions. In 1918, Fisher produced the paper "The Correlation between Relatives on the Supposition of Mendelian Inheritance," which showed how the continuous variation measured by the biometricians could be the result of the action of many discrete genetic loci. In this and subsequent papers culminating in his 1930 book *The Genetical Theory of Natural Selection*, Fisher was able to show how Mendelian genetics was, contrary to the thinking of many early geneticists, completely consistent with the idea of evolution driven by natural selection. During the 1920s, a series of papers by Haldane applied mathematical analysis to real-world examples of natural selection such as the evolution of industrial melanism in peppered moths. Haldane established that natural selection could work in the real world at a faster rate than even Fisher had assumed.

Sewall Wright focused on combinations of genes that interacted as complexes, and the effects of inbreeding on small relatively isolated populations, which could exhibit genetic drift. In a 1932 paper, he introduced the concept of an adaptive landscape in which phenomena such as cross breeding and genetic drift in small populations could push them away from adaptive peaks, which would in turn allow natural selection to push them towards new adaptive peaks. Wright's model would appeal to field naturalists such as Theodosius Dobzhansky and Ernst Mayr who were becoming aware of the importance of geographical isolation in real world populations. The work of Fisher, Haldane and Wright founded the discipline of population genetics. This is the precursor of the modern synthesis, which is an even broader coalition of ideas. One limitation of the modern synthesis version of population genetics is that it treats one gene locus at a time, neglecting genetic linkage and resulting linkage disequilibrium between loci.

The Modern Synthesis

Theodosius Dobzhansky, an emigrant from the Soviet Union to the United States, who had been a postdoctoral worker in Morgan's fruit fly lab, was one of the first to apply genetics to natural populations. He worked mostly with *Drosophila pseudoobscura*. He says pointedly: "Russia has a variety of climates from the Arctic to sub-tropical... Exclusively laboratory workers who neither possess nor wish to have any knowledge of living beings in nature were and are in a minority." Not surprisingly, there were other Russian geneticists with similar ideas, though for some time their work was known to only a few in the West. His 1937 work *Genetics and the Origin of Species* was a key step in bridging the gap between population geneticists and field naturalists. It presented the conclusions reached by Fisher, Haldane, and especially Wright in their highly mathematical papers in a form that was easily accessible to others. It also emphasized that real world populations had far more genetic variability than the early population geneticists had assumed in their models, and that genetically distinct sub-populations were important. Dobzhansky argued that natural selection worked to maintain genetic diversity as well as driving change. Dobzhansky had been influenced by his exposure in the 1920s to the work of a Russian geneticist Sergei Chetverikov who had looked at the role of recessive genes in maintaining a reservoir of genetic variability in a population before his work was shut down by the rise of Lysenkoism in the Soviet Union.

E. B. Ford's work complemented that of Dobzhansky. It was as a result of Ford's work, as well as his own, that Dobzhansky changed the emphasis in the third edition of his famous text from drift to selection. Ford was an experimental naturalist who wanted to test natural selection in nature. He virtually invented the field of research known as ecological genetics. His work on natural selection in wild populations of butterflies and moths was the first to show that predictions made by R. A. Fisher were correct. He was the first to describe and define genetic polymorphism, and to predict that human blood group polymorphisms might be maintained in the population by providing some protection against disease.

Ernst Mayr's key contribution to the synthesis was *Systematics and the Origin of Species*, published in 1942. Mayr emphasized the importance of allopatric speciation, where geographically isolated sub-populations diverge so far that reproductive isolation occurs. He was skeptical of the reality of sympatric speciation believing that geographical isolation was a prerequisite for building up intrinsic (reproductive) isolating mechanisms. Mayr also introduced the biological species concept that defined a species as a group of interbreeding or potentially interbreeding populations that were reproductively isolated from all other populations. Before he left Germany for the United States in 1930, Mayr had been influenced by the work of German biologist Bernhard Rensch. In the 1920s Rensch, who like Mayr did field work in Indonesia, analyzed the geographic distribution of polytypic species and complexes of closely related species paying particular attention to how variations between different populations correlated with local environmental factors such as differences in climate. In 1947, Rensch published *Neuere Probleme der Abstammungslehre. Die transspezifische Evolution* (1959 English translation of 2nd edition: *Evolution Above the Species Level*). This looked at how the same evolutionary mechanisms involved in speciation might be extended to explain the origins of the differences between the higher level taxa. His writings contributed to the rapid acceptance of the synthesis in Germany.

George Gaylord Simpson was responsible for showing that the modern synthesis was compatible with paleontology in his book *Tempo and Mode in Evolution* published in 1944. Simpson's work was crucial because so many paleontologists had disagreed, in some cases vigorously, with the idea that natural selection was the main mechanism of evolution. It showed that the trends of linear progression (in for example the evolution of the horse) that earlier paleontologists had used as support for neo-Lamarckism and orthogenesis did not hold up under careful examination. Instead the fossil record was consistent with the irregular, branching, and non-directional pattern predicted by the modern synthesis.

The botanist G. Ledyard Stebbins was another major contributor to the synthesis. His major work, *Variation and Evolution in Plants*, was published in 1950. It extended the synthesis to encompass botany including the important effects of hybridization and polyploidy in some kinds of plants.

Further Advances

The modern evolutionary synthesis continued to be developed and refined after the initial establishment in the 1930s and 1940s. The work of W. D. Hamilton, George C. Williams, John Maynard Smith and others led to the development of a gene-centered view of evolution in the 1960s. The synthesis as it exists now has extended the scope of the Darwinian idea of natural selection to include subsequent scientific discoveries and concepts unknown to Darwin, such as DNA and genetics, which allow rigorous, in many cases mathematical, analyses of phenomena such as kin selection, altruism, and speciation.

John Maynard Smith.

In *The Selfish Gene* (1976), author Richard Dawkins argues the gene is the only true unit of selection. (Dawkins also attempts to apply evolutionary theory to non-biological entities, such as cultural memes, imagined to be subject to selective forces analogous to those affecting biological entities.)

Richard Dawkins.

Others, such as Stephen Jay Gould, reject the notion that genetic entities are subject to anything other than genetic or chemical forces, (as well as the idea evolution acts on "populations" per se), reasserting the centrality of the individual organism as the true unit of selection, whose specific phenotype is directly subject to evolutionary pressures.

In 1972, the notion of gradualism in evolution was challenged by a theory of punctuated equilibrium put forward by Gould and Niles Eldredge, proposing evolutionary changes could occur in relatively rapid spurts, when selective pressures were heightened, punctuating long periods of morphological stability, as well-adapted organisms coped successfully in their respective environments.

Discovery in the 1980s of Hox genes and regulators conserved across multiple phyletic divisions began the process of addressing basic theoretical problems relating to gradualism, incremental change, and sources of novelty in evolution. Suddenly, evolutionary theorists could answer the charge that spontaneous random mutations should result overwhelmingly in deleterious changes to a fragile, monolithic genome: Mutations in homeobox regulation could safely—yet dramatically—alter morphology at a high level, without damaging coding for specific organs or tissues.

This, in turn, provided the means to model hypothetical genomic changes expressed in the phenotypes of long-extinct species, like the Late Devonian "fish with hands," *Tiktaalik*, discovered in 2004.

As these discoveries suggest, the synthesis continues to undergo regular review, drawing on insights offered by both new biotechnologies and new paleontological discoveries.

After the Synthesis

There are a number of discoveries in earth sciences and biology which have arisen since the synthesis. Listed here are some of those topics which are relevant to the evolutionary synthesis, and which seem soundly based.

Understanding of Earth History

The Earth is the stage on which the evolutionary play is performed. Darwin studied evolution in the context of Charles Lyell's geology, but our present understanding of Earth history includes some critical advances made during the last half-century.

- The age of the Earth has been revised upwards. It is now estimated to be approximately 4.54 billion years old, nearly one-third of the age of the universe. The Phanerozoic (current eon) only occupies the last one-ninth of this period of time.
- The triumph of Alfred Wegener's idea of continental drift came around 1960. The key principle of plate tectonics is that the lithosphere exists as separate and distinct tectonic plates, which ride on the fluid-like (visco-elastic solid) asthenosphere. This discovery provides a unifying theory for geology, linking phenomena such as volcanoes, earthquakes, orogeny, and providing data for many paleogeographical questions. One major question is still unclear: when did plate tectonics begin?
- Understanding of the evolution of the atmosphere of Earth has progressed. The substitution of oxygen for carbon dioxide (CO_2) in the atmosphere, which occurred in the Proterozoic, caused probably by cyanobacteria in the form of stromatolites, caused changes leading to the evolution of aerobic organisms.
- The identification of the first generally accepted fossils of microbial life was made by geologists. These rocks have been dated as about 3.465 billion years ago. Charles Doolittle Walcott was the first geologist to identify Precambrian fossil bacteria from microscopic examination of thin rock slices. He also thought stromatolites were organic in origin. His ideas were not accepted at the time, but may now be appreciated as great discoveries.
- Information about paleoclimates is increasingly available, and being used in paleontology. One example: the discovery of massive ice ages in the Proterozoic, following the great re-

duction of CO_2 in the atmosphere. These ice ages were immensely long, and led to a crash in microflora.

- Catastrophism and extinction events. A partial reintegration of catastrophism has occurred, and the importance of mass extinctions in large-scale evolution is now apparent. Extinction events disturb relationships between many forms of life and may remove dominant forms and release a flow of adaptive radiation amongst groups that remain. Causes include meteorite strikes (Cretaceous–Paleogene extinction event; Late Devonian extinction); flood basalt provinces (Deccan Traps at Cretaceous–Paleogene boundary; Siberian Traps at P–Tr); and other less dramatic processes.

Conclusion: Our present knowledge of earth history strongly suggests that large-scale geophysical events influenced macroevolution and megaevolution. These terms refer to evolution above the species level, including such events as mass extinctions, adaptive radiation, and the major transitions in evolution.

Symbiotic Origin of Eukaryotic Cell Structures

Once symbiosis was discovered in lichen and in plant roots (rhizobia in root nodules) in the 19th century, the idea arose that the process might have occurred more widely, and might be important in evolution. Heinrich Anton de Bary invented the concept of *symbiosis*; several Russian biologists promoted the idea; Edmund Beecher Wilson mentioned it in his book *The Cell in Development and Heredity* (1925); as did Ivan Wallin in his *Symbionticism and the Origin of Species* (1927); and there was a brief mention by Julian Huxley in 1930; all in vain because sufficient evidence was lacking. Symbiosis as a major evolutionary force was not discussed at all in the evolutionary synthesis.

The role of symbiosis in cell evolution was revived partly by Joshua Lederberg, and finally brought to light by Lynn Margulis in a series of papers and books. Some organelles are recognized as being of microbial origin: mitochondria and chloroplasts definitely, cilia, flagella and centrioles possibly, and perhaps the nuclear membrane and much of the chromosome structure as well. What is now clear is that the evolution of eukaryote cells is either caused by, or at least profoundly influenced by, symbiosis with bacterial and archaean cells in the Proterozoic.

The origin of the eukaryote cell by symbiosis in several stages was not part of the evolutionary synthesis. It is, at least on first sight, an example of megaevolution by big jumps. However, what symbiosis provided was a copious supply of heritable variation from microorganisms, which was fine-tuned over a long period to produce the cell structure we see today. This part of the process is consistent with evolution by natural selection.

Trees of Life

The ability to analyse sequence in macromolecules (protein, DNA, RNA) provides evidence of descent, and permits us to work out genealogical trees covering the whole of life, since now there are data on every major group of living organisms. This project, begun in a tentative way in the 1960s, has become a search for the universal tree or the universal ancestor, a phrase of Carl Woese. The tree that results has some unusual features, especially in its roots. There are two domains of prokaryotes: bacteria and archaea, both of which contributed genetic material to the eukaryotes, mainly by means of symbiosis. Also, since bacteria can pass genetic material to other bacteria, their relationships look more like a web than a tree. Once eukaryotes were established, their sexual reproduction produced the traditional branching tree-like pattern, the only diagram Darwin published in *On the Origin of Species*. The last universal ancestor (LUA) would be a prokaryotic cell before the split between the bacteria and archaea. LUA is defined as most recent organism from which all organisms now living on Earth descend (some 3.5 to 3.8 billion years ago, in the Archean eon).

This technique may be used to clarify relationships within any group of related organisms. It is now a standard procedure, and examples are published regularly. April 2008 sees the publication of a tree covering all the animal phyla, derived from sequences from 150 genes in 77 taxa.

Early attempts to identify relationships between major groups were made in the 19th century by Ernst Haeckel, and by comparative anatomists such as Thomas Henry Huxley and E. Ray Lankester. Enthusiasm waned: it was often difficult to find evidence to adjudicate between different opinions. Perhaps for that reason, the evolutionary synthesis paid surprisingly little attention to this activity. It is certainly a lively field of research today.

Evolutionary Developmental Biology

What once was called embryology played a modest role in the evolutionary synthesis, mostly about evolution by changes in developmental timing (allometry and heterochrony). Man himself was, according to Louis Bolk, a typical case of evolution by retention of juvenile characteristics (neoteny). He listed many characters where "Man, in his bodily development, is a primate foetus that has become sexually mature." Unfortunately, his interpretation of these ideas was non-Darwinian, but his list of characters is both interesting and convincing.

Evolutionary developmental biology springs from clear proof that development is closely controlled by special genetic systems, and the hope that comparison of these systems will tell us much about the evolutionary history of different groups. In a series of experiments with the fruit fly *Drosophila*, Edward B. Lewis was able to identify a complex of genes whose proteins bind to the cis-regulatory regions of target genes. The latter then activate or repress systems of cellular processes that accomplish the final development of the organism. Furthermore, the sequence of these control genes show *co-linearity*: the order of the loci in the chromosome parallels the order in which the loci are expressed along the anterior-posterior axis of the body. Not only that, but this cluster of master control genes programs the development of all higher organisms. Each of the genes contains a homeobox, a remarkably conserved DNA sequence. This suggests the complex itself arose by gene duplication. In his 1995 Nobel lecture, Lewis said "Ultimately, comparisons of the HOM-C throughout the animal kingdom should provide a picture of how the organisms, as well as the genes of the HOM-C, have evolved."

The term *deep homology* was coined to describe the common origin of genetic regulatory apparatus used to build morphologically and phylogenetically disparate animal features. It applies when a complex genetic regulatory system is inherited from a common ancestor, as it is in the evolution of vertebrate and invertebrate eyes. The phenomenon is implicated in many cases of parallel evolution.

A great deal of evolution may take place by changes in the control of development. This may be relevant to punctuated equilibrium theory, for in development a few changes to the control system could make a significant difference to the adult organism. An example is the giant panda, whose place in the Carnivora was long uncertain. Apparently, the giant panda's evolution required the change of only a few genetic messages (5 or 6 perhaps), yet the phenotypic and lifestyle change from a standard bear is considerable. The transition could therefore be effected relatively swiftly.

Fossil Discoveries

Since the late 20th century, there have been excavations in parts of the world which had scarcely been investigated before. Also, there is fresh appreciation of fossils discovered in the 19th century, but then denied or deprecated: the classic example is the Ediacaran biota from the immediate Precambrian, after the Cryogenian period. These soft-bodied fossils are the first record of multicellular life. The interpretation of this fauna is still in flux.

Many outstanding discoveries have been made, and some of these have implications for evolutionary theory. The discovery of feathered dinosaurs and early birds from the Early Cretaceous of Liaoning Province in northeast China have convinced most paleontologists that birds did evolve from coelurosaurian theropod dinosaurs. Less well known, but perhaps of equal evolutionary significance, are the studies on early insect flight, on stem tetrapods from the Late Devonian, and the early stages of whale evolution.

Recent work has shed light on the evolution of flatfish (order Pleuronectiformes), such as plaice, sole, turbot and halibut. Flatfish are interesting because they are one of the few vertebrate groups with external asymmetry. Their young are perfectly symmetrical, but the head is remodelled during a metamorphosis, which entails the migration of one eye to the other side, close to the other eye. Some species have both eyes on the left (turbot), some on the right (halibut, sole); all living and fossil flatfish to date show an 'eyed' side and a 'blind' side. The lack of an intermediate condition in living and fossil flatfish species had led to debate about the origin of such a striking adaptation. The case was considered by Jean-Baptiste Lamarck, who thought flatfish precursors would have lived in shallow water for a long period, and by Charles Darwin, who predicted a gradual migration of the eye, mirroring the metamorphosis of the living forms. Darwin's long-time critic St. George Mivart thought that the intermediate stages could have no selective value, and in the 6th edition of *On the Origin of Species*, Darwin made a concession to the possibility of acquired traits. Many years later the geneticist Richard Goldschmidt put the case forward as an example of evolution by saltation, bypassing intermediate forms.

A recent examination of two fossil species from the Eocene has provided the first clear picture of flatfish evolution. The discovery of stem flatfish with incomplete orbital migration refutes Goldschmidt's ideas, and demonstrates that "the assembly of the flatfish bodyplan occurred in a gradual, stepwise fashion." There are no grounds for thinking that incomplete orbital migration was maladaptive, because stem forms with this condition ranged over two geological stages, and are found in localities which also yield flatfish with the full cranial asymmetry. The evolution of flatfish falls squarely within the evolutionary synthesis.

Horizontal Gene Transfer

Horizontal gene transfer (HGT) (or *lateral gene transfer*) is any process in which an organism gets genetic material from another organism without being the offspring of that organism.

Most thinking in genetics has focused on vertical transfer, but there is a growing awareness that horizontal gene transfer is a significant phenomenon. Amongst single-celled organisms it may be the dominant form of genetic transfer. Artificial horizontal gene transfer is a form of genetic engineering.

Aaron O. Richardson and Jeffrey D. Palmer state: "Horizontal gene transfer (HGT) has played a major role in bacterial evolution and is fairly common in certain unicellular eukaryotes. However, the prevalence and importance of HGT in the evolution of multicellular eukaryotes remain unclear."

The bacterial means of HGT are:

- Transformation, the genetic alteration of a cell resulting from the introduction, uptake and expression of foreign genetic material (DNA or RNA).
- Transduction, the process in which bacterial DNA is moved from one bacterium to another by a bacterial virus (a bacteriophage, or 'phage').
- Bacterial conjugation, a process in which a bacterial cell transfers genetic material to another cell by cell-to-cell contact.
- Gene transfer agent (GTA) is a virus-like element which contains random pieces of the host chromosome. They are found in most members of the Alphaproteobacteria order Rhodobacterales. They are encoded by the host genome. GTAs transfer DNA so frequently that they may have an important role in evolution. A 2010 report found that genes for antibiotic resistance could be transferred by engineering GTAs in the laboratory.

Some examples of HGT in Metazoa are now known. Genes in bdelloid rotifers have been found which appear to have originated in bacteria, fungi, and plants. This suggests they arrived by horizontal gene transfer. The capture and use of exogenous (~foreign) genes may represent an important force in bdelloid evolution. The team led by Matthew S. Meselson at Harvard University has also shown that, despite the lack of sexual reproduction, bdelloid rotifers do engage in genetic (DNA) transfer within a species or clade. The method used is not known at present.

Adaptation

In biology, an *adaptation*, also called an *adaptive trait*, is a trait with a current functional role in the life of an organism that is maintained and evolved by means of natural selection. Adaptation refers to both the current state of being adapted and to the dynamic evolutionary process that leads to the adaptation. Adaptations enhance the fitness and survival of individuals. Organisms face a succession of environmental challenges as they grow and develop and are equipped with an adaptive plasticity as the phenotype of traits develop in response to the imposed conditions. The developmental norm of reaction for any given trait is essential to the correction of adaptation as it affords a kind of biological insurance or resilience to varying environments.

General Principles

Adaptation is, first of all, a *process*, rather than a part of a body. An internal parasite (such as a liver fluke) can illustrate the distinction: such a parasite may have a very simple bodily structure, but nevertheless the organism is highly adapted to its specific environment. From this we see that adaptation is not just a matter of visible traits: in such parasites critical adaptations take place in the life cycle, which is often quite complex. However, as a practical term, adaptation is often used for the *product*: those features of a species which result from the process. Many aspects of an animal or plant can be correctly called adaptations, though there are always some features whose function is in doubt. By using the term *adaptation* for the evolutionary *process*, and *adaptive trait* for the bodily part or function (the product), the two senses of the word may be distinguished.

Adaptation is one of the two main processes that explain the diverse species we see in biology, such as the different species of Darwin's finches. The other is speciation (species-splitting or cladogenesis), caused by geographical isolation or some other mechanism. A favorite example used today to study the interplay of adaptation and speciation is the evolution of cichlid fish in African lakes, where the question of reproductive isolation is much more complex.

Adaptation is not always a simple matter, where the ideal phenotype evolves for a given external environment. An organism must be viable at all stages of its development and at all stages of its evolution. This places *constraints* on the evolution of development, behavior and structure of organisms. The main constraint, over which there has been much debate, is the requirement that each genetic and phenotypic change during evolution should be relatively small, because developmental systems are so complex and interlinked. However, it is not clear what "relatively small" should mean, for example polyploidy in plants is a reasonably common large genetic change. The origin of eukaryotic symbiosis is a more dramatic example.

All adaptations help organisms survive in their ecological niches. These adaptive traits may be structural, behavioral or physiological. Structural adaptations are physical features of an organism (shape, body covering, armament; and also the internal organization). Behavioral adaptations are composed of inherited behavior chains and/or the ability to learn: behaviors may be inherited in detail (instincts), or a capacity for learning may be inherited. Examples: searching for food, mating, vocalizations. Physiological adaptations permit the organism to perform special functions (for instance, making venom, secreting slime, phototropism); but also more general functions such as growth and development, temperature regulation, ionic balance and other aspects of homeostasis. Adaptation, then, affects all aspects of the life of an organism.

Definitions

The following definitions are mainly due to Theodosius Dobzhansky.

1. *Adaptation* is the evolutionary process whereby an organism becomes better able to live in its habitat or habitats.
2. *Adaptedness* is the state of being adapted: the degree to which an organism is able to live and reproduce in a given set of habitats.
3. An *adaptive trait* is an aspect of the developmental pattern of the organism which enables or enhances the probability of that organism surviving and reproducing.

Adaptedness and Fitness

From the above definitions, it is clear that there is a relationship between adaptedness and *fitness* (a key population genetics concept). Differences in fitness between genotypes predict the rate of evolution by natural selection. Natural selection changes the relative frequencies of alternative phenotypes, insofar as they are heritable. Although the two are connected, the one does not imply the other: a phenotype with high adaptedness may not have high fitness. Dobzhansky mentioned the example of the Californian redwood, which is highly adapted, but a relict species in danger of extinction. Elliott Sober commented that adaptation was a retrospective concept since it implied something about the history of a trait, whereas fitness predicts a trait's future.

1. Relative fitness. The average contribution to the next generation by a genotype or a class of genotypes, relative to the contributions of other genotypes in the population. This is also known as *Darwinian fitness*, *selection coefficient*, and other terms.
2. Absolute fitness. The absolute contribution to the next generation by a genotype or a class of genotypes. Also known as the Malthusian parameter when applied to the population as a whole.
3. Adaptedness. The extent to which a phenotype fits its local ecological niche. This can sometimes be tested through a reciprocal transplant.

Brief History

Adaptation is a fact of life that has been accepted by many of the great thinkers who have tackled the world of living organisms. It is their explanations of how adaptation arises that separates these thinkers. A few of the most significant ideas:

- Empedocles did not believe that adaptation required a final cause (~ purpose), but "came about naturally, since such things survived." Aristotle, however, did believe in final causes.

Jean-Baptiste Lamarck (1744–1829).

- In natural theology, adaptation was interpreted as the work of a deity, even as evidence for the existence of God. William Paley believed that organisms were perfectly adapted to the lives they lead, an argument that shadowed Gottfried Wilhelm Leibniz, who had argued that God had brought about "the best of all possible worlds." Voltaire's Dr. Pangloss is a parody of this optimistic idea, and David Hume also argued against design. The *Bridgewater Treatises* are a product of natural theology, though some of the authors managed to present their work in a fairly neutral manner. The series was lampooned by Robert Knox, who held quasi-evolutionary views, as the *Bilgewater Treatises*. Charles Darwin broke with the tradition by emphasising the flaws and limitations which occurred in the animal and plant worlds.

- Lamarckism is a proto-evolutionary hypothesis of the inheritance of acquired characteristics, whose main purpose is to explain adaptations by natural means. Jean-Baptiste Lamarck proposed a tendency for organisms to become more complex, moving up a ladder of progress, plus "the influence of circumstances," usually expressed as *use and disuse*. His evolutionary ideas, and those of Étienne Geoffroy Saint-Hilaire, fail because they cannot be reconciled with heredity. This was known even before Gregor Mendel by medical men interested in human races (William Charles Wells, William Lawrence), and especially by August Weismann.

Many other students of natural history, such as Buffon, accepted adaptation, and some also accepted evolution, without voicing their opinions as to the mechanism. This illustrates the real merit of Darwin and Alfred Russel Wallace, and secondary figures such as Henry Walter Bates, for putting forward a mechanism whose significance had only been glimpsed previously. A century later, experimental field studies and breeding experiments by people such as E. B. Ford and Dobzhansky produced evidence that natural selection was not only the 'engine' behind adaptation, but was a much stronger force than had previously been thought.

Types of Adaptations

Adaptation is the heart and soul of evolution.

—Niles Eldredge, Reinventing Darwin: The Great Debate at the High Table of Evolutionary Theory

Changes in Habitat

Before Charles Darwin, adaptation was seen as a fixed relationship between an organism and its habitat. It was not appreciated that as the climate changed, so did the habitat; and as the habitat changed, so did the biota. Also, habitats are subject to changes in their biota: for example, invasions of species from other areas. The relative numbers of species in a given habitat are always changing. Change is the rule, though much depends on the speed and degree of the change.

When the habitat changes, three main things may happen to a resident population: habitat tracking, genetic change or extinction. In fact, all three things may occur in sequence. Of these three effects, only genetic change brings about adaptation.

Habitat Tracking

When a habitat changes, the most common thing to happen is that the resident population moves to another locale which suits it; this is the typical response of flying insects or oceanic organisms, who have wide (though not unlimited) opportunity for movement. This common response is called *habitat tracking.* It is one explanation put forward for the periods of apparent stasis in the fossil record (the punctuated equilibrium theory).

Genetic Change

Genetic change is what occurs in a population when natural selection acts on the genetic variability of the population; moreover, some mutations may create genetic variation that will lead to differing characteristics of offspring and hence abet adaptation. The first pathways of enzyme-based metabolism may have been parts of purine nucleotide metabolism, with previous metabolic pathways being part of the ancient RNA world. By this means, the population adapts genetically to its circumstances. Genetic changes may result in visible structures, or may adjust physiological activity in a way that suits the habitat.

It is now clear that habitats and biota do frequently change. Therefore, it follows that the process of adaptation is never finally complete. Over time, it may happen that the environment changes little, and the species comes to fit its surroundings better and better. On the other hand, it may happen that changes in the environment occur relatively rapidly, and then the species becomes less and less well adapted. Seen like this, adaptation is a genetic *tracking process*, which goes on all the time to some extent, but especially when the population cannot or does not move to another, less hostile area. Also, to a greater or lesser extent, the process affects every species in a particular ecosystem.

Leigh Van Valen thought that even in a stable environment, competing species had to constantly adapt to maintain their relative standing. This became known as the Red Queen hypothesis. One of the manifestations of the Red Queen dynamics can be seen in host-parasite interaction.

Intimate Relationships: Co-Adaptations

In coevolution, where the existence of one species is tightly over bound up with the life of another species, new or 'improved' adaptations which occur in one species are often followed by the appearance and spread of corresponding features in the other species. There are many examples of this; the idea emphasises that the life and death of living things is intimately connected, not just with the physical environment, but with the life of other species. These relationships are intrinsically dynamic, and may continue on a trajectory for millions of years, as has the relationship between flowering plants and insects (pollination).

Pollinator constancy: these honey bees selectively visit flowers from only one species, as can be seen by the colour of the pollen in their baskets:

The gut contents, wing structures, and mouthpart morphologies of fossilized beetles and flies suggest that they acted as early pollinators. The association between beetles and angiosperms during the Early Cretaceous period led to parallel radiations of angiosperms and insects into the Late Cretaceous. The evolution of nectaries in Late Cretaceous flowers signals the beginning of the mutualism between hymenopterans and angiosperms.

Mimicry

A and B show real wasps; the rest are mimics: three hoverflies and one beetle.

Bates' work on Amazonian butterflies led him to develop the first scientific account of mimicry, especially the kind of mimicry which bears his name: Batesian mimicry. This is the mimicry by a palatable species of an unpalatable or noxious species. A common example seen in temperate gardens is the hoverfly, many of which–though bearing no sting–mimic the warning colouration of hymenoptera (wasps and bees). Such mimicry does not need to be perfect to improve the survival of the palatable species.

Bates, Wallace and Fritz Müller believed that Batesian and Müllerian mimicry provided evidence for the action of natural selection, a view which is now standard amongst biologists. All aspects of this situation can be, and have been, the subject of research. Field and experimental work on these ideas continues to this day; the topic connects strongly to speciation, genetics and development.

The Basic Machinery: Internal Adaptations

There are some important adaptations to do with the overall coordination of the systems in the body. Such adaptations may have significant consequences. Examples, in vertebrates, would be temperature regulation, or improvements in brain function, or an effective immune system. An example in plants would be the development of the reproductive system in flowering plants. Such adaptations may make the clade (monophyletic group) more viable in a wide range of habitats. The acquisition of such major adaptations has often served as the spark for adaptive radiation, and huge success over long periods of time for a whole group of animals or plants.

Compromise and Conflict Between Adaptations

It is a profound truth that Nature does not know best; that genetical evolution... is a story of waste, makeshift, compromise and blunder.

—Peter Medawar, The Future of Man

All adaptations have a downside: horse legs are great for running on grass, but they can't scratch their backs; mammals' hair helps temperature, but offers a niche for ectoparasites; the only flying penguins do is under water. Adaptations serving different functions may be mutually destructive. Compromise and makeshift occur widely, not perfection. Selection pressures pull in different directions, and the adaptation that results is some kind of compromise.

Since the phenotype as a whole is the target of selection, it is impossible to improve simultaneously all aspects of the phenotype to the same degree.

—Ernst Mayr, The Growth of Biological Thought: Diversity, Evolution, and Inheritance

Consider the antlers of the Irish elk, (often supposed to be far too large; in deer antler size has an allometric relationship to body size). Obviously, antlers serve positively for defence against predators, and to score victories in the annual rut. But they are costly in terms of resource. Their size during the last glacial period presumably depended on the relative gain and loss of reproductive capacity in the population of elks during that time. Another example: camouflage to avoid detection is destroyed when vivid colors are displayed at mating time. Here the risk to life is counterbalanced by the necessity for reproduction.

Stream-dwelling salamanders, such as Caucasian salamander or Gold-striped salamander have very slender, long bodies, perfectly adapted to life at the banks of fast small rivers and mountain brooks. Elongated body protects their larvae from being washed out by current. However, elongated body increases risk of desiccation and decreases dispersal ability of the salamanders; it also negatively affects their fecundity. As a result, fire salamander, less perfectly adapted to the mountain brook habitats, is in general more successful, have a higher fecundity and broader geographic range.

An Indian peacock's trainin full display.

The peacock's ornamental train (grown anew in time for each mating season) is a famous adaptation. It must reduce his maneuverability and flight, and is hugely conspicuous; also, its growth costs food resources. Darwin's explanation of its advantage was in terms of sexual selection: "This depends on the advantage which certain individuals have over other individuals of the same sex and species, in exclusive relation to reproduction." The kind of sexual selection represented by the peacock is called 'mate choice,' with an implication that the process selects the more fit over the less fit, and so has survival value. The recognition of sexual selection was for a long time in abeyance, but has been rehabilitated. In practice, the Indian peafowl (*Pavo cristatus*) is a successful species, with a large natural range in India, so the overall outcome of their mating system is quite viable.

The conflict between the size of the human foetal brain at birth, (which cannot be larger than about 400 cm^3, else it will not get through the mother's pelvis) and the size needed for an adult brain (about 1400 cm^3), means the brain of a newborn child is quite immature. The most vital things in human life (locomotion, speech) just have to wait while the brain grows and matures. That is the result of the birth compromise. Much of the problem comes from our upright bipedal stance, without which our pelvis could be shaped more suitably for birth. Neanderthals had a similar problem.

As another example, the long neck of a giraffe is a burden and a blessing. The neck of a giraffe can be up to 2 m (6 ft 7 in) in length. This neck can be used for inter-species competition or for foraging on tall trees where shorter herbivores cannot reach. However, as previously stated, there is always a trade-off. This long neck is heavy and it adds to the body mass of a giraffe, so the giraffe needs an abundance of nutrition to provide for this costly adaptation.

Shifts in Function

Adaptation and function are two aspects of one problem.

—Julian Huxley, Evolution: The Modern Synthesis

Pre-Adaptations

This occurs when a species or population has characteristics which (by chance) are suited for conditions which have not yet arisen. For example, the polyploid cordgrass *Spartina townsendii* is better adapted than either of its parent species to their own habitat of saline marsh and mud-flats. White Leghorn chicken are markedly more resistant to vitamin B_1 deficiency than other breeds. On a plentiful diet there is no difference, but on a restricted diet this preadaptation could be decisive.

Pre-adaptation may occur because a natural population carries a huge quantity of genetic variability. In diploid eukaryotes, this is a consequence of the system of sexual reproduction, where mutant alleles get partially shielded, for example, by the selective advantage of heterozygotes. Microorganisms, with their huge populations, also carry a great deal of genetic variability.

The first experimental evidence of the pre-adaptive nature of genetic variants in microorganisms was provided by Salvador Luria and Max Delbrück who developed Fluctuation Test, a method to show the random fluctuation of pre-existing genetic changes that conferred resistance to bacteriophage in the bacterium *Escherichia coli.*

Co-option of Existing Traits: Exaptation

The classic example is the ear ossicles of mammals, which we know from paleontological and embryological studies originated in the upper and lower jaws and the hyoid bone of their synapsid ancestors, and further back still were part of the gill arches of early fish. We owe this esoteric knowledge to the comparative anatomists, who, a century ago, were at the cutting edge of evolutionary studies. The word *exaptation* was coined to cover these shifts in function, which are surprisingly common in evolutionary history. The origin of wings from feathers that were originally used for temperature regulation is a more recent discovery.

Related Issues

Non-adaptive Traits

Some traits do not appear to be adaptive, that is, they appear to have a neutral or even deleterious effect on fitness in the current environment. Because genes have pleiotropic effects, not all traits may be functional (i.e. spandrels). Alternatively, a trait may have been adaptive at some point in an organism's evolutionary history, but a change in habitats caused what used to be an adaptation to become unnecessary or even a hindrance (maladaptations). Such adaptations are termed vestigial.

Vestigial Organs

Many organisms have vestigial organs, which are the remnants of fully functional structures in their ancestors. As a result of changes in lifestyle the organs became redundant, and are either not functional or reduced in functionality. With the loss of function goes the loss of positive selection, and the subsequent accumulation of deleterious mutations. Since any structure represents some kind of cost to the general economy of the body, an advantage may accrue from their elimination once they are not functional. Examples: wisdom teeth in humans; the loss of pigment and functional eyes in cave fauna; the loss of structure in endoparasites.

Fitness Landscapes

Sewall Wright proposed that populations occupy *adaptive peaks* on a fitness landscape. In order to evolve to another, higher peak, a population would first have to pass through a valley of maladaptive intermediate stages. A given population might be "trapped" on a peak that is not optimally adapted.

Extinction

If a population cannot move or change sufficiently to preserve its long-term viability, then obviously, it will become extinct, at least in that locale. The species may or may not survive in other locales. Species extinction occurs when the death rate over the entire species exceeds the birth rate for a long enough period for the species to disappear. It was an observation of Van Valen that groups of species tend to have a characteristic and fairly regular rate of extinction.

Coextinction

Just as we have co-adaptation, there is also coextinction. Coextinction refers to the loss of a species due to the extinction of another; for example, the extinction of parasitic insects following the loss of their hosts. Coextinction can also occur when a flowering plant loses its pollinator, or through the disruption of a food chain. Ecologist Lian Pin Koh and colleagues discuss coextinction, stating, "Species coextinction is a manifestation of the interconnectedness of organisms in complex ecosystems. . . . While coextinction may not be the most important cause of species extinctions, it is certainly an insidious one."

Flexibility, Acclimatization, Learning

Generalists, such as birds, are sometimes able to adapt to urban areas.

Flexibility deals with the relative capacity of an organism to maintain themselves in different habitats: their degree of specialization. *Acclimatization* is a term used for automatic physiological adjustments during life; *learning* is the term used for improvement in behavioral performance during life. In biology these terms are preferred, not adaptation, for changes during life which improve the performance of individuals. These adjustments are not inherited genetically by the next generation.

Adaptation, on the other hand, occurs over many generations; it is a gradual process caused by natural selection which changes the genetic make-up of a population so the collective performs better in its niche.

Flexibility

Populations differ in their phenotypic plasticity, which is the ability of an organism with a given genotype to change its phenotype in response to changes in its habitat, or to move to a different habitat.

To a greater or lesser extent, all living things can adjust to circumstances. The degree of flexibility is inherited, and varies to some extent between individuals. A highly specialized animal or plant lives only in a well-defined habitat, eats a specific type of food, and cannot survive if its needs are not met. Many herbivores are like this; extreme examples are koalas which depend on *eucalyptus*, and giant pandas which require bamboo. A generalist, on the other hand, eats a range of food, and can survive in many different conditions. Examples are humans, rats, crabs and many carnivores. The *tendency* to behave in a specialized or exploratory manner is inherited—it is an adaptation.

Rather different is *developmental flexibility:* "An animal or plant is developmentally flexible if when it is raised in or transferred to new conditions, it changes in structure so that it is better fitted to survive in the new environment," writes evolutionary biologist John Maynard Smith. Once again, there are huge differences between species, and the *capacities* to be flexible are inherited.

Acclimatization

If humans move to a higher altitude, respiration and physical exertion become a problem, but after spending time in high altitude conditions they *acclimatize* to the pressure by increasing production of Red blood cells. The *ability* to acclimatize is an adaptation, but not the acclimatization itself. Fecundity goes down, but deaths from some tropical diseases also goes down.

Over a longer period of time, some people will reproduce better at these high altitudes than others. They will contribute more heavily to later generations. Gradually the whole population becomes adapted to the new conditions. This we know takes place, because the performance of long-term communities at higher altitude is significantly better than the performance of new arrivals, even when the new arrivals have had time to make physiological adjustments.

Some kinds of acclimatization happen so rapidly that they are better called reflexes. The rapid colour changes in some flatfish, cephalopods, chameleons are examples.

Learning

Social learning is supreme for humans, and is possible for quite a few mammals and birds: of course, that does not involve genetic transmission except to the extent that the capacities are inherited. Similarly, the *capacity to learn* is an inherited adaptation, but not what is learnt; the capacity for human speech is inherited, but not the details of language.

Diversity of Genome DNAs

A large diversity of genome DNAs in a species is the basis for species' adaptation and for species' differentiation. A great number of individuals are needed for carrying the great number of different genome DNAs. According to the Misrepair-accumulation aging theory, Misrepair mechanism is important in maintaining the sufficient number of individuals in a species. Misrepair is a way of repair for increasing the surviving chance of an organism when it has severe injuries. Without Misrepairs, no individual could survive to reproduction age. Thus Misrepair mechanism is an essential mechanism for the survival of a species and for maintaining the number of individuals. Although individuals die from aging, genome DNAs are being recopied and transmitted by individuals generation by generation. In addition, the DNA Misrepairs in germ cells contribute also to the diversity of genome DNAs.

Function and Teleonomy

Adaptation raises some issues concerning how biologists use key terms such as *function*.

Function

To say something has a function is to say something about what it does for the organism, obviously. It also says something about its history: how it has come about. A heart pumps blood: that is its function. It also emits sound, which is just an ancillary side-effect. That is not its function. The heart has a history (which may be well or poorly understood), and that history is about how natural selection formed and maintained the heart as a pump. Every aspect of an organism that has a function has a history. Now, an adaptation must have a functional history: therefore we expect it must have undergone selection caused by relative survival in its habitat. It would be quite wrong to use the word adaptation about a trait which arose as a by-product.

It is widely regarded as unprofessional for a biologist to say something like "A wing is for flying," although that is their normal function. A biologist would be conscious that sometime in the remote past feathers on a small dinosaur had the function of retaining heat, and that later many wings were not used for flying (e.g. penguins, ostriches). So, the biologist would rather say that the wings on a bird or an insect usually had the *function* of aiding flight. That would carry the connotation of being an adaptation with a history of evolution by natural selection.

Teleonomy

Teleonomy is a term invented to describe the study of goal-directed functions which are not guided by the conscious forethought of man or any supernatural entity. It is contrasted with Aristotle's teleology, which has connotations of intention, purpose and foresight. Evolution is teleonomic; *adaptation hoards hindsight rather than foresight.* The following is a definition for its use in biology:

> Teleonomy: The hypothesis that adaptations arise without the existence of a prior purpose, but by chance may change the fitness of an organism.

The term may have been suggested by Colin Pittendrigh in 1958; it grew out of cybernetics and self-organising systems. Ernst Mayr, George C. Williams and Jacques Monod picked up the term and used it in evolutionary biology.

Philosophers of science have discussed the concept. Ernest Nagel analysed goal-directedness in biology; and David Hull commented on the use of teleology in biology:

> ...Haldane can be found remarking, 'Teleology is like a mistress to a biologist: he cannot live without her but he's unwilling to be seen with her in public.' Today the mistress has become a lawfully wedded wife. Biologists no longer feel obligated to apologize for their use of teleological language; they flaunt it. The only concession which they make to its disreputable past is to rename it 'teleonomy'.

Genetic Variation

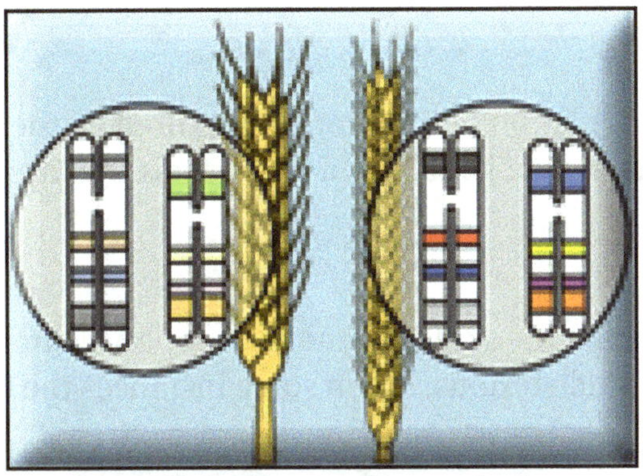

Darwin's finches or Galapagos finches.

Genetic variation is a fact that a biological system – individual and population – is different over space. It is the base of the Genetic variability of different biological systems in space.

Genetic variation is based on the variation in alleles of genes in a gene pool. It occurs both within and among populations, supported by individual carriers of the variant genes. Genetic variation is brought about by random mutation, which is a permanent change in the chemical structure of a gene.

Among Individuals within a Population

Genetic variation among individuals within a population can be identified at a variety of levels. It is possible to identify genetic variation from observations of phenotypic variation in either quantitative traits (traits that vary continuously and are coded for by many genes (e.g., leg length in dogs)) or discrete traits (traits that fall into discrete categories and are coded for by one or a few genes (e.g., white, pink, red petal color in certain flowers)).

Genetic variation can also be identified by examining variation at the level of enzymes using the process of protein electrophoresis. Polymorphic genes have more than one allele at each locus. Half of the genes that code for enzymes in insects and plants may be polymorphic, whereas polymorphisms are less common in vertebrates.

Ultimately, genetic variation is caused by variation in the order of bases in the nucleotides in genes. New technology now allows scientists to directly sequence DNA which has identified even more genetic variation than was previously detected by protein electrophoresis. Examination of DNA has shown genetic variation in both coding regions and in the non-coding intron region of genes.

Genetic variation will result in phenotypic variation if variation in the order of nucleotides in the DNA sequence results in a difference in the order of amino acids in proteins coded by that DNA sequence, and if the resultant differences in amino acid sequence influence the shape, and thus the function of the enzyme.

Between Populations

Geographic variation in genes often occurs among populations living in different locations. Geographic variation may be due to differences in selective pressures or to genetic drift.

Measurement

Genetic variation within a population is commonly measured as the percentage of gene loci that are polymorphic or the percentage of gene loci in individuals that are heterozygous.

Sources

Random mutations are the ultimate source of genetic variation. Mutations are likely to be rare and most mutations are neutral or deleterious, but in some instances the new alleles can be favored by natural selection.

A range of variability in the mussel *Donax variabilis*.

Variability of walnuts.

Polyploidy is an example of chromosomal mutation. Polyploidy is a condition wherein organisms have three or more sets of genetic variation (3n or more). The mutation is started off by a parent, as the parent mates the offspring now has a chance to receive that mutation trait also. Now when that mutated offspring is ready to mate they now have the chance of passing on that trait to their offspring. This process begins the first generation of mutated offspring.

Crossing over and random segregation during meiosis can result in the production of new alleles or new combinations of alleles. Furthermore, random fertilization also contributes to variation.

Variation and recombination can be facilitated by transposable and transposed genetic elements, commonly known as endogenous retroviruses, LINEs, SINEs, etc.

For a given genome of a multicellular organism, genetic variation may be acquired in somatic cells or inherited through the germline.

Forms

Genetic variation can be divided into different forms according to the size and type of genomic variation underpinning genetic change. These include:

- Small-scale sequence variation (< 1Kbp):
    - Substitution
    - Indels
- Large-scale Structural variation (>1Kbp):
    - Copy number variation:
        - Copy number loss
        - Copy number gain
    - Rearrangement:
        - Translocation
        - Inversion
        - Segmental acquired uniparental disomy
- Numerical variation (whole chromosomes or genomes):
    - Polyploidy
    - Aneuploidy

Maintenance in Populations

A variety of factors maintain genetic variation in populations. Potentially harmful recessive alleles can be hidden from selection in the heterozygous individuals in populations of diploid organisms (recessive alleles are only expressed in the less common homozygous individuals). Natural selection can also maintain genetic variation in balanced polymorphisms. Balanced polymorphisms may occur when heterozygotes are favored or when selection is frequency dependent.

References

- Borenstein, Seth (13 November 2013). "Oldest fossil found: Meet your microbial mom". Excite. Yonkers, NY: Mindspark Interactive Network. Associated Press. Retrieved 01, May 2020
- Ridley, Mark. "Speciation - What is the role of reinforcement in speciation?". Retrieved 2015-09-07. Adapted from Evolution (2004), 3rd edition (Malden, MA: Blackwell Publishing), ISBN 978-1-4051-0345-9
- Pinker, Steven (June 18, 2012). "The False Allure of Group Selection". edge.org. Edge Foundation, Inc. Retrieved 05, June 2020
- Cavali-Sforza L. L., Bodmer W. F. (1999): The genetics of human populations. Dover Publications, Inc., Mineola, New York, ISBN 0-486-40693-8
- Rabajante, J; et al. (2016). "Host-parasite Red Queen dynamics with phase-locked rare genotypes". Science Advances. 2: e1501548. doi:10.1126/sciadv.1501548. ISSN 2375-2548
- Minkel, J. R. (September 8, 2006). "Wandering Fly Gene Supports New Model of Speciation". Scientific American. Stuttgart: Georg von Holtzbrinck Publishing Group. ISSN 0036-8733. Retrieved 21, June 2020
- Wang-Michelitsch, Jicun; Michelitsch, Thomas M. (2015). "Aging as a process of accumulation of Misrepairs". arXiv:1503.07163 [q-bio.TO]
- Wang-Michelitsch, Jicun; Michelitsch, Thomas M. (2015). "Misrepair mechanism: a mechanism essential for individual adaptation, species adaptation and species evolution". arXiv:1505.03900 [q-bio.TO]

Chapter 5

The Process of Natural Selection

The biological process whereby organisms get adapted to the given environment in order to survive and reproduce is known as natural selection. This chapter contains topics like heritable traits, sexual selection, stabilizing selection, directional selection, negative selection, gene selection, etc. The aim of this chapter is to explore the field of natural selection.

Natural Selection in action

Natural selection is the engine that drives evolution. The organisms best suited to survive in their particular circumstances have a greater chance of passing their traits on to the next generation. But plants and animals interact in very complex ways with other organisms and their environment.

Natural Selection Examples

Here are examples of natural selection:

- In a habitat there are red bugs and green bugs. The birds prefer the taste of the red bugs, so soon there are many green bugs and few red bugs. The green bugs reproduce and make more green bugs and eventually there are no more red bugs.
- In an ecosystem, some giraffes have long necks and others have short ones. If something caused low-lying shrubs to die out, the giraffes with short necks would not get enough food. After a few generations, all the giraffes would have long necks.
- A species of rats live in a certain type of tree with the branches evenly spaced. Smaller rats could not reach from branch to branch and larger rats would break the branches and fall. Soon, all rats were just the right size for the tree branches.

- Deer mice that migrated to the sand hills of Nebraska changed from dark brown to light brown to better hide from predators in the sand.
- Galapagos finches all have different types of beaks. During drought, the finches with the larger beaks survived better than those with smaller beaks. During rainy times, more small seeds were produced and the finches with smaller beaks fared better.
- In one ecosystem, lizards that had long legs could climb better to avoid floods and reach food.
- Insects become resistant to pesticides very quickly, sometime in one generation. If an insect is resistant to the chemical, most of the offspring will also be resistant. Considering that insect generations can be a matter of weeks, insects in an area can become immune to a chemical within months.
- The bacterium Pseudomonas metabolizes nylon; but, when a certain type of this bacterium that did not eat nylon was placed in an environment where nylon was the only food, the bacterium evolved until it ate the nylon.
- Peacock females pick their mate according to the male's tail. The ones with the largest and brightest tails mate more often. Today, males that do not have bright feathers are very rare.
- The peppered moth used to be a light color with black spots. When the atmosphere in London became filled with soot because of the Industrial Revolution, the white trees became darker and light colored moths were eaten by birds more readily. Within months, moths became darker and lighter moths were rare. After the Industrial Revolution, light colored moths were found in greater numbers.
- Rat snakes are all very similar except in coloring. Some are striped and some are green, black, and orange. This is due to living in many different types of terrain and adapting to the environment.
- Warrior ants have a chemical signal that tells other ants in the family not to attack. Some have adapted and learned to imitate the chemical signal from other colonies so they can invade and take over another colony and the workers will never know.
- Sharks are colored white on the underside and blue or gray on the top. This is their camouflage as the top blends with the water color to someone looking down into the water and the bottom blends with the light coming through the water from above.
- Bacteria have become resistant to antibiotics and this can happen very quickly, since bacteria can produce several generations within one day. The strongest bacteria are the last to die and the antibiotics sometimes do not kill all of them.

- The field mustard plant survived a drought in southern California because of genetic changes that made it produces flowers earlier. Plants that survive a drought have to have a short growing season and the field mustard evolved quickly.
- Because of its long body, the moray eel's mouth did not produce enough suction to catch prey; so, it adapted and grew a second set of jaws and teeth.
- Humans have evolved and are still evolving. There are people more resistant to malaria that live in Africa.
- Humans don't become lactose intolerant as some species do. This is thought to be because of the domestication of cattle.
- Early humans survived because of a certain hand shape that could either toss a spear or throw a rock the best.
- In the 18th century, researchers found in an island community where there was a genetic disposition to mate earlier and have larger families. Over 140 years, the age at first reproduction dropped from 26 to 22 and this was due to natural selection.

Charles Darwin coined the term "natural selection. Darwin's concept of natural selection was based on several key observations:

- Traits are often heritable. In living organisms, many characteristics are inherited, or passed from parent to offspring. (Darwin knew this was the case, even though he did not know that traits were inherited via genes).

- More offspring are produced than can survive. Organisms are capable of producing more offspring than their environments can support. Thus, there is competition for limited resources in each generation.

- Offspring vary in their heritable traits. The offspring in any generation will be slightly different from one another in their traits (color, size, shape, etc.), and many of these features will be heritable.

Based on these simple observations, Darwin concluded the following:

- In a population, some individuals will have inherited traits that help them survive and reproduce (given the conditions of the environment, such as the predators and food sources present). The individuals with the helpful traits will leave more offspring in the next generation than their peers, since the traits make them more effective at surviving and reproducing.
- Because the helpful traits are heritable, and because organisms with these traits leave more offspring, the traits will tend to become more common (present in a larger fraction of the population) in the next generation.
- Over generations, the population will become adapted to its environment (as individuals with traits helpful in that environment have consistently greater reproductive success than their peers).

Darwin's model of evolution by natural selection allowed him to explain the patterns he had seen during his travels. For instance, if the Galápagos finch species shared a common ancestor, it made sense that they should broadly resemble one another (and mainland finches, who likely shared that common ancestor). If groups of finches had been isolated on separate islands for many generations, however, each group would have been exposed to a different environment in which different heritable traits might have been favored, such as different sizes and shapes of beaks for using different food sources. These factors could have led to the formation of distinct species on each island.

Mechanism

Heritable Variation, Differential Reproduction

Natural variation occurs among the individuals of any population of organisms. Some differences may improve an individual's chances of surviving and reproducing such that its lifetime reproductive rate is increased, which means that it leaves more offspring. If the traits that give these individuals a reproductive advantage are also heritable, that is, passed from parent to offspring, then there will be differential reproduction, that is, a slightly higher proportion of fast rabbits or efficient algae in the next generation. Even if the reproductive advantage is very slight, over many generations any advantageous heritable trait becomes dominant in the population. In this way the natural environment of an organism "selects for" traits that confer a reproductive advantage, causing evolutionary change, as Darwin described. This gives the appearance of purpose, but in natural selection there is no intentional choice. Artificial selection is purposive where natural selection is not, though biologists often use teleological language to describe it.

During the industrial revolution, pollution killed many lichens, leaving tree trunks dark. A dark (melanic) morph of the moth largely replaced the formerly usual light morph (both shown here). Since the moths are subject to predation by birds hunting by sight, the color change offers better camouflage against the changed background, suggesting natural selection at work.

The peppered moth exists in both light and dark colours in Great Britain, but during the industrial revolution, many of the trees on which the moths rested became blackened by soot, giving the dark-coloured moths an advantage in hiding from predators. This gave dark-coloured moths a better chance of surviving to produce dark-coloured offspring, and in just fifty years from the first dark moth being caught, nearly all of the moths in industrial Manchester were dark. The balance was reversed by the effect of the Clean Air Act 1956, and the dark moths became rare again, demonstrating the influence of natural selection on peppered moth evolution. A recent study, using image analysis and avian vision models, shows that pale individuals more closely match lichen backgrounds than dark morphs and for the first time quantifies the camouflage of moths to predation risk.

Fitness

The concept of fitness is central to natural selection. In broad terms, individuals that are more "fit" have better potential for survival, as in the well-known phrase "survival of the fittest", but the precise meaning of the term is much more subtle. Modern evolutionary theory defines fitness not by how long an organism lives, but by how successful it is at reproducing. If an organism lives half as long as others of its species, but has twice as many offspring surviving to adulthood, its genes become more common in the adult population of the next generation. Though natural selection acts on individuals, the effects of chance mean that fitness can only really be defined "on average" for the individuals within a population. The fitness of a particular genotype corresponds to the average effect on all individuals with that genotype.

Competition

In biology, competition is an interaction between organisms in which the fitness of one is lowered by the presence of another. This may be because both rely on a limited supply of a resource such as food, water, or territory. Competition may be within or between species, and may be direct or indirect. Species less suited to compete should in theory either adapt or die out, since competition plays a powerful role in natural selection, but according to the "room to roam" theory it may be less important than expansion among larger clades.

Competition is modeled by *r/K* selection theory, which is based on Robert MacArthur and E. O. Wilson's work on island biogeography. In this theory, selective pressures drive evolution in one of two stereotyped directions: *r*- or *K*-selection. These terms, *r* and *K*, can be illustrated in a logistic model of population dynamics:

$$\frac{dN}{dt} = rN\left(1 - \frac{N}{K}\right)$$

Where *r* is the growth rate of the population (*N*), and *K* is the carrying capacity of its local environmental setting. Typically, *r*-selected species exploit empty niches, and produce many offspring, each with a relatively low probability of surviving to adulthood. In contrast, *K*-selected species are strong competitors in crowded niches, and invest more heavily in much fewer offspring, each with a relatively high probability of surviving to adulthood.

Types of Selection

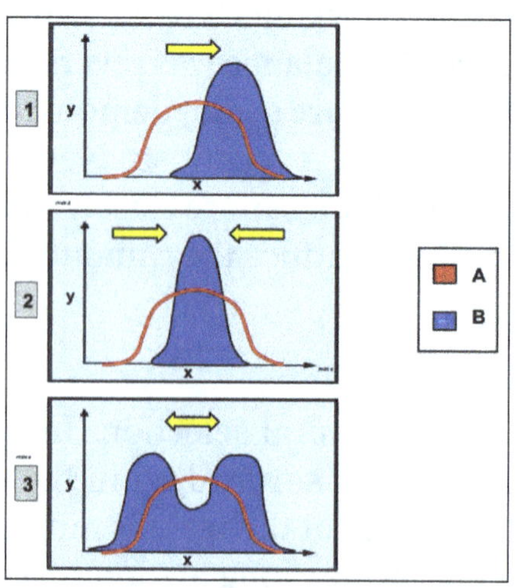

1: directional selection: a single extreme phenotype favoured. 2, stabilizing selection: intermediate favoured over extremes. 3: disruptive selection: extremes favoured over intermediate. X-axis: phenotypic trait, Y-axis: number of organisms, Group A: original population, Group B: after selection.

Natural selection can act on any heritable phenotypic trait, and selective pressure can be produced by any aspect of the environment, including sexual selection and competition with members of the same or other species. However, this does not imply that natural selection is always directional and results in adaptive evolution; natural selection often results in the maintenance of the status quo by eliminating less fit variants.

Selection can be classified in several different ways, such as by its effect on a trait, on genetic diversity, by the life cycle stage where it acts, by the unit of selection, or by the resource being competed for.

Selection has different effects on traits. Stabilizing selection acts to hold a trait at a stable optimum, and in the simplest case all deviations from this optimum are selectively disadvantageous. Directional selection favours extreme values of a trait. The uncommon disruptive selection also acts during transition periods when the current mode is sub-optimal, but alters the trait in more than one direction. In particular, if the trait is quantitative and univariate then both higher and lower trait levels are favoured. Disruptive selection can be a precursor to speciation.

Alternatively, selection can be divided according to its effect on genetic diversity. Purifying or negative selection acts to remove genetic variation from the population (and is opposed by *de novo* mutation, which introduces new variation.In contrast, balancing selection acts to maintain genetic variation in a population, even in the absence of *de novo* mutation, by negative frequency-dependent selection. One mechanism for this is heterozygote advantage, where individuals with two different alleles have a selective advantage over individuals with just one allele. The polymorphism at the human ABO blood group locus has been explained in this way.

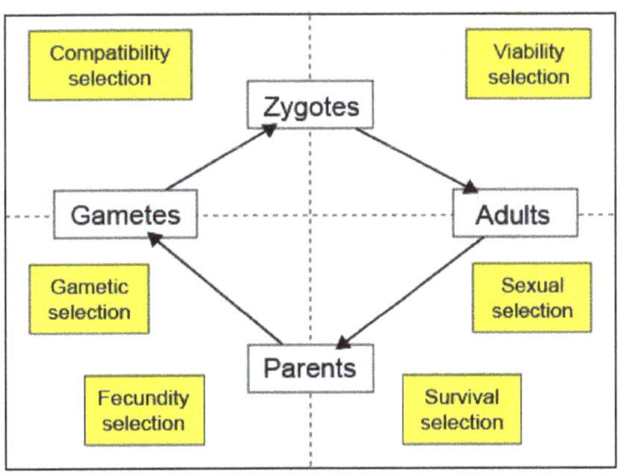

Different types of selection act at each life cycle stage of a sexually reproducing organism.

Another option is to classify selection by the life cycle stage at which it acts. Some biologists recognize just two types: viability (or survival) selection, which acts to increase an organism's probability of survival, and fecundity (or fertility or reproductive) selection, which acts to increase the rate of reproduction, given survival. Others split the life cycle into further components of selection. Thus viability and survival selection may be defined separately and respectively as acting to improve the probability of survival before and after reproductive age is reached, while fecundity selection may be split into additional sub-components including sexual selection, gametic selection, acting on gamete survival, and compatibility selection, acting on zygote formation.

Selection can also be classified by the level or unit of selection. Individual selection acts on the individual, in the sense that adaptations are "for" the benefit of the individual, and result from selection among individuals. Gene selection acts directly at the level of the gene. In kin selection and intragenomic conflict, gene-level selection provides a more apt explanation of the underlying process. Group selection, if it occurs, acts on groups of organisms, on the assumption that groups replicate and mutate in an analogous way to genes and individuals. There is an ongoing debate over the degree to which group selection occurs in nature.

Finally, selection can be classified according to the resource being competed for. Sexual selection results from competition for mates. Sexual selection typically proceeds via fecundity selection, sometimes at the expense of viability. Ecological selection is natural selection via any means other than sexual selection, such as kin selection, competition, and infanticide. Following Darwin, natural selection is sometimes defined as ecological selection, in which case sexual selection is considered a separate mechanism.

Sexual Selection

The peacock's elaborate plumage is mentioned by Darwin as an example of sexual selection, and is a classic example of Fisherian runaway, driven to its conspicuous size and coloration through mate choice by females over many generations.

Sexual selection as first articulated by Darwin (using the example of the peacock's tail) refers specifically to competition for mates, which can be *intrasexual*, between individuals of the same sex, that is male–male competition, or *intersexual*, where one gender chooses mates, most often with males displaying and females choosing. However, in some species, mate choice is primarily by males, as in some fishes of the family Syngnathidae.

Phenotypic traits can be displayed in one sex and desired in the other sex, causing a positive feedback loop called a Fisherian runaway, for example, the extravagant plumage of some male birds such as the peacock. An alternate theory proposed by the same Ronald Fisher in 1930 is the sexy son hypothesis, that mothers want promiscuous sons to give them large numbers of grandchildren and so choose promiscuous fathers for their children. Aggression between members of the same sex is sometimes associated with very distinctive features, such as the antlers of stags, which are used in combat with other stags. More generally, intrasexual selection is often associated with sexual dimorphism, including differences in body size between males and females of a species.

Natural Selection in Action

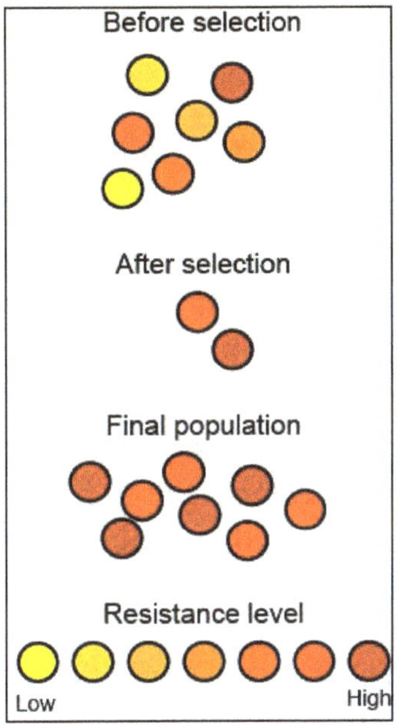

Selection in action: resistance to antibiotics grows though the survival of individuals less affected by the antibiotic. Their offspring inherit the resistance.

Natural selection is seen in action in the development of antibiotic resistance in microorganisms. Since the discovery of penicillinin 1928, antibiotics have been used to fight bacterial diseases. The widespread misuse of antibiotics has selected for microbial resistance to antibiotics in clinical use, to the point that the methicillin-resistant Staphylococcus aureus (MRSA) has been described as a "superbug" because of the threat it poses to health and its relative invulnerability to existing drugs. Response strategies typically include the use of different, stronger antibiotics; however, new strains of MRSA have recently emerged that are resistant even to these drugs. This is an evolutionary arms race, in which bacteria develop strains less susceptible to antibiotics, while medical researchers attempt to develop new antibiotics that can kill them. A similar situation occurs with pesticide resistance in plants and insects. Arms races are not necessarily induced by man; a well-documented example involves the spread of a gene in the butterfly Hypolimnas bolina suppressing male-killing activity by *Wolbachia* bacteria parasites on the island of Samoa, where the spread of the gene is known to have occurred over a period of just five years.

Evolution by Means of Natural Selection

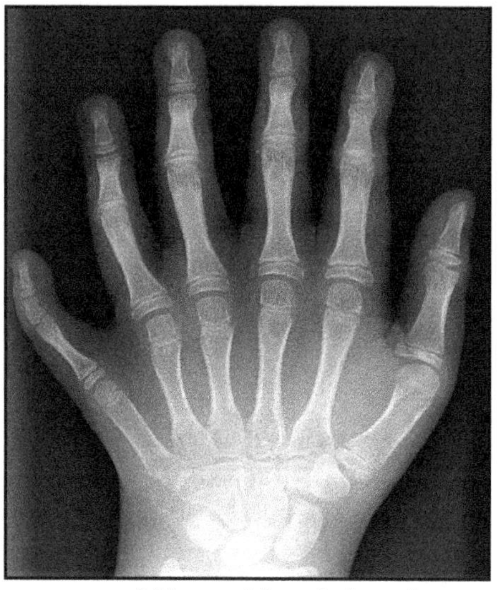

X-ray of the left hand of a ten-year-old boy with polydactyly, caused by a mutant *Hox* gene.

A prerequisite for natural selection to result in adaptive evolution, novel traits and speciation is the presence of heritable genetic variation that results in fitness differences. Genetic variation is the result of mutations, genetic recombinations and alterations in the karyotype (the number, shape, size and internal arrangement of the chromosomes). Any of these changes might have an effect that is highly advantageous or highly disadvantageous, but large effects are rare. In the past, most changes in the genetic material were considered neutral or close to neutral because they occurred in noncoding DNA or resulted in a synonymous substitution. However, many mutations in non-coding DNA have deleterious effects. Although both mutation rates and average fitness effects of mutations are dependent on the organism, a majority of mutations in humans are slightly deleterious.

Some mutations occur in "toolkit" or regulatory genes. Changes in these often have large effects on the phenotype of the individual because they regulate the function of many other genes. Most, but not all, mutations in regulatory genes result in non-viable embryos. Some nonlethal regulatory mutations occur in HOX genes in humans, which can result in a cervical rib or polydactyl, an increase in the number of fingers or toes. When such mutations result in a higher fitness, natural selection favours these phenotypes and the novel trait spreads in the population. Established traits are not immutable; traits that have high fitness in one environmental context may be much less fit if environmental conditions change. In the absence of natural selection to preserve such a trait, it becomes more variable and deteriorate over time, possibly resulting in a vestigial manifestation of the trait, also called evolutionary baggage. In many circumstances, the apparently vestigial structure may retain a limited functionality, or may be co-opted for other advantageous traits in a phenomenon known as preadaptation. A famous example of a vestigial structure, the eye of the blind mole-rat, is believed to retain function in photoperiod perception.

Speciation

Speciation requires a degree of reproductive isolation—that is, a reduction in gene flow. However, it is intrinsic to the concept of a species that hybrids are selected against, opposing the evolution of reproductive isolation, a problem that was recognized by Darwin. The problem does not occur in allopatric speciation with geographically separated populations, which can diverge with different sets of mutations. E. B. Poulton realized in 1903 that reproductive isolation could evolve through divergence, if each lineage acquired a different, incompatible allele of the same gene. Selection against the heterozygote would then directly create reproductive isolation, leading to the Bateson–Dobzhansky–Muller model, further elaborated by H. Allen Orr and Sergey Gavrilets. With reinforcement, however, natural selection can favor an increase in pre-zygotic isolation, influencing the process of speciation directly.

Genetic Basis

Genotype and Phenotype

Natural selection acts on an organism's phenotype, or physical characteristics. Phenotype is determined by an organism's genetic make-up (genotype) and the environment in which the organism lives. When different organisms in a population possess different versions of a gene for a certain trait, each of these versions is known as an allele. It is this genetic variation that underlies differences in phenotype. An example is the ABO blood type antigens in humans, where three alleles govern the phenotype.

Some traits are governed by only a single gene, but most traits are influenced by the interactions of many genes. A variation in one of the many genes that contributes to a trait may have only a small effect on the phenotype; together, these genes can produce a continuum of possible phenotypic values.

Directionality of Selection

When some component of a trait is heritable, selection alters the frequencies of the different alleles, or variants of the gene that produces the variants of the trait. Selection can be divided into three classes, on the basis of its effect on allele frequencies: directional, stabilizing, and purifying selection. Directional selection occurs when an allele has a greater fitness than others, so that it increases in frequency, gaining an increasing share in the population. This process can continue until the allele is fixed and the entire population shares the fitter phenotype. Far more common is stabilizing selection, which lowers the frequency of alleles that have a deleterious effect on the phenotype – that is, produce organisms of lower fitness. This process can continue until the allele is eliminated from the population. Purifying selection conserves functional genetic features, such as protein-coding genes or regulatory sequences, over time by selective pressure against deleterious variants.

Some forms of balancing selection do not result in fixation, but maintain an allele at intermediate frequencies in a population. This can occur in diploid species (with pairs of chromosomes) when heterozygous individuals (with just one copy of the allele) have a higher fitness than homozygous individuals (with two copies). This is called heterozygote advantage or over-dominance, of which the best-known example is the resistance to malaria in humans heterozygous for sickle-cell anaemia. Maintenance of allelic variation can also occur through disruptive or diversifying selection, which favours genotypes that depart from the average in either direction (that is, the opposite of over-dominance), and can result in a bimodal distribution of trait values. Finally, balancing selection can occur through frequency-dependent selection, where the fitness of one particular phenotype depends on the distribution of other phenotypes in the population. The principles of game theory have been applied to understand the fitness distributions in these situations, particularly in the study of kin selection and the evolution of reciprocal altruism.

Selection, Genetic Variation, and Drift

A portion of all genetic variation is functionally neutral, producing no phenotypic effect or significant difference in fitness. Motoo Kimura's neutral theory of molecular evolution by genetic drift proposes that this variation accounts for a large fraction of observed genetic diversity. Neutral events can radically reduce genetic variation through population bottlenecks. which among other things can cause the founder effect in initially small new populations. When genetic variation does not result in differences in fitness, selection cannot directly affect the frequency of such variation. As a result, the genetic variation at those sites is higher than at sites where variation does influence fitness. However, after a period with no new mutations, the genetic variation at these sites is eliminated due to genetic drift. Natural selection reduces genetic variation by eliminating maladapted individuals, and consequently the mutations that caused the maladaptation. At the same time, new mutations occur, resulting in a mutation–selection balance. The exact outcome of the two processes depends both on the rate at which new mutations occur and on the strength of the natural selection, which is a function of how unfavorable the mutation proves to be.

Genetic linkage occurs when the loci of two alleles are in close proximity on a chromosome. During the formation of gametes, recombination reshuffles the alleles. The chance that such a reshuffle occurs between two alleles is inversely related to the distance between them. Selective sweeps occur when an allele becomes more common in a population as a result of positive selection. As the prevalence of one allele increases, closely linked alleles can also become more common by "genetic hitchhiking", whether they are neutral or even slightly deleterious. A strong selective sweep results in a region of the genome where the positively selected haplotype (the allele and its neighbours) are in essence the only ones that exist in the population. Selective sweeps can be detected by measuring linkage disequilibrium, or whether a given haplotype is overrepresented in the population. Since a selective sweep also results in selection of neighbouring alleles, the presence of a block of strong linkage disequilibrium might indicate a 'recent' selective sweep near the center of the block.

Background selection is the opposite of a selective sweep. If a specific site experiences strong and persistent purifying selection, linked variation tends to be weeded out along with it, producing a region in the genome of low overall variability. Because background selection is a result of deleterious new mutations, which can occur randomly in any haplotype, it does not produce clear blocks of linkage disequilibrium, although with low recombination it can still lead to slightly negative linkage disequilibrium overall.

Impact

Darwin's ideas, along with those of Adam Smith and Karl Marx, had a profound influence on 19th century thought, including his radical claim that "elaborately constructed forms, so different from each other, and dependent on each other in so complex a manner" evolved from the simplest forms of life by a few simple principles. This inspired some of Darwin's most ardent supporters—and provoked the strongest opposition. Natural selection had the power, according to Stephen Jay Gould, to "dethrone some of the deepest and most traditional comforts of Western thought", such as the belief that humans have a special place in the world.

In the words of the philosopher Daniel Dennett, "Darwin's dangerous idea" of evolution by natural selection is a "universal acid," which cannot be kept restricted to any vessel or container, as it soon leaks out, working its way into ever-wider surroundings. Thus, in the last decades, the concept of natural selection has spread from evolutionary biology to other disciplines, including evolutionary computation, quantum Darwinism, evolutionary economics, evolutionary epistemology, evolutionary psychology, and cosmological natural selection. This unlimited applicability has been called universal Darwinism.

Origin of Life

How life originated from inorganic matter remains an unresolved problem in biology. One prominent hypothesis is that life first appeared in the form of short self-replicating RNA polymers. On this view, life may have come into existence when RNA chains first experienced the basic conditions, as conceived by Charles Darwin, for natural selection to operate. These conditions are: heritability, variation of type, and competition for limited resources. The fitness of an early RNA replicator would likely have been a function of adaptive capacities that were intrinsic (i.e., determined by the nucleotide sequence) and the availability of resources. The three primary adaptive capacities could logically have been:

(1) The capacity to replicate with moderate fidelity (giving rise to both heritability and variation of type),

(2) The capacity to avoid decay,

(3) The capacity to acquire and process resources.

These capacities would have been determined initially by the folded configurations (including those configurations with ribozyme activity) of the RNA replicators that, in turn, would have been encoded in their individual nucleotide sequences.

Cell and Molecular Biology

In 1881, the embryologist Wilhelm Roux published *Der Kampf der Theile im Organismus* (*The Struggle of Parts in the Organism*) in which he suggested that the development of an organism results from a Darwinian competition between the parts of the embryo, occurring at all levels, from molecules to organs. In recent years, a modern version of this theory has been proposed by Jean-Jacques Kupiec. According to this cellular Darwinism, random variation at the molecular level generates diversity in cell types whereas cell interactions impose a characteristic order on the developing embryo.

Social and Psychological Theory

The social implications of the theory of evolution by natural selection also became the source of continuing controversy. Friedrich Engels, a German political philosopher and co-originator of the ideology of communism, wrote in 1872 that "Darwin did not know what a bitter satire he wrote on mankind, and especially on his countrymen, when he showed that free competition, the struggle for existence, which the economists celebrate as the highest historical achievement, is the normal state of the *animal kingdom*." Herbert Spencer and the eugenics advocate Francis Galton's interpretation of natural selection as necessarily progressive, leading to supposed advances in intelligence and civilization, became a justification for colonialism, eugenics, and social Darwinism. For example, in 1940, Konrad Lorenz, in writings that he subsequently disowned, used the theory as a justification for policies of the Nazi state. He wrote "selection for toughness, heroism, and social utility must be accomplished by some human institution, if mankind, in default of selective factors, is not to be ruined by domestication-induced degeneracy. The racial idea as the basis of our state has already accomplished much in this respect." Others have developed ideas that human societies and culture evolve by mechanisms analogous to those that apply to evolution of species.

More recently work among anthropologists and a psychologist has led to the development of sociobiology and later of evolutionary psychology, a field that attempts to explain features of human psychology in terms of adaptation to the ancestral environment. The most prominent example of evolutionary psychology, notably advanced in the early work of Noam Chomsky and later by Steven Pinker, is the hypothesis that the human brain has adapted to acquire the grammatical rules of natural language. Other aspects of human behavior and social structures, from specific cultural norms such as incest avoidance to broader patterns such as gender roles, have been hypothesized to have similar origins as adaptations to the early environment in which modern humans evolved. By analogy to the action of natural selection on genes, the concept of memes—"units of cultural transmission," or culture's equivalents of genes undergoing selection and recombination—has arisen, first described in this form by Richard Dawkins in 1976 and subsequently expanded upon by philosophers such as Daniel Dennettas explanations for complex cultural activities, including human consciousness.

Information and Systems Theory

In 1922, Alfred J. Lotka proposed that natural selection might be understood as a physical principle that could be described in terms of the use of energy by a system, a concept later developed by Howard T. Odum as the maximum power principle in thermodynamics, whereby evolutionary systems with selective advantage maximise the rate of useful energy transformation.

The principles of natural selection have inspired a variety of computational techniques, such as "soft" artificial life, that simulate selective processes and can be highly efficient in 'adapting' entities to an environment defined by a specified fitness function. For example, a class of heuristic optimisation algorithms known as genetic algorithms, pioneered by John Henry Holland in the 1970s and expanded upon by David E. Goldberg, identify optimal solutions by simulated reproduction and mutation of a population of solutions defined by an initial probability distribution. Such algorithms are particularly useful when applied to problems whose energy landscape is very rough or has many local minima.

In Fiction

Darwinian evolution by natural selection is pervasive in literature, whether taken optimistically in terms of how humanity may evolve towards perfection, or pessimistically in terms of the dire consequences of the interaction of human nature and the struggle for survival. Among major responses is Samuel Butler's 1872 pessimistic *Erewhon* ("nowhere", written backwards). In 1893 H. G. Wells imagined "The Man of the Year Million", transformed by natural selection into a being with a huge head and eyes, and shrunken body.

Heritable Traits

A heritable quantitative trait is a measurable phenotype that depends on the cumulative actions of many genes, and the environment. These traits can vary among individuals, over a range, to produce a continuous distribution of phenotypes.

Examples of Heritable Traits

Earlobe Attachment

If earlobes hang free, they are detached. If they connect directly to the sides of the head, they are attached. Earlobe attachment is a continuous trait: while most earlobes can be neatly categorized as attached or unattached, some are in-between.

Although some sources say that this trait is controlled by a single gene, with unattached earlobes being dominant over attached earlobes, no published studies support this view. Earlobe attachment and shape are inherited, but it is likely that many genes contribute to this trait. As such, its pattern of inheritance is difficult to predict.

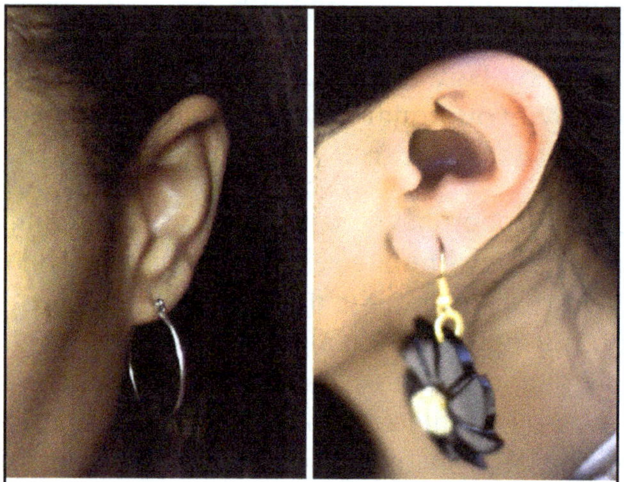

Attached earlobe. Unattached earlobe.

Tongue Rolling

Some people can curl up the sides of their tongue to form a tube shape. In 1940, Alfred Sturtevant observed that about 70% of people of European ancestry could roll their tongues and the remaining 30% could not.

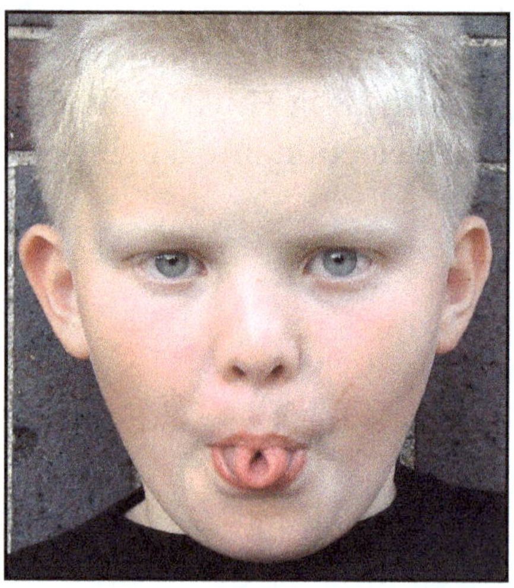

Many sources state that tongue rolling is controlled by a single gene. However, as Sturtevant observed, people can learn to roll their tongue as they get older, suggesting that environmental factors—not just genes—influence the trait. Consistent with this view, just 70% of identical twins share the trait (if tongue rolling were influenced only by genes, then 100% of identical traits would share the trait).

Dimples

Dimples are small, natural indentations on the cheeks. They can appear on one or both sides, and they often change with age. Some people are born with dimples that disappear when they're adults; others develop dimples later in childhood.

Dimples are highly heritable, meaning that people who have dimples tend to have children with dimples—but not always. Because their inheritance isn't completely predictable, dimples are considered an "irregular" dominant trait. Having dimples is probably controlled mainly by one gene but also influenced by other genes.

Handedness

Handedness describes our preference for using either our left or right hand for activities such as writing and throwing a ball. Overall, about 10% of people are left-handed, but the number varies among cultures from 0.5% to 24%.

Some have reported that handedness is controlled by just one or two genes, but this is not the case. Multiple studies present evidence that handedness is controlled by many genes—at least 30 and as many as 100—each with a small effect; many are linked to brain development. Environment also plays an important role: some cultures actively discourage left-handedness.

Freckles

Freckles are small, concentrated spots of a skin pigment called melanin. Most fair-skinned, red-haired people have them.

Freckles are controlled primarily by the MC1R gene. Freckles show a dominant inheritance pattern: parents who have freckles tend to have children with freckles.

Variations, also called alleles, of MC1R control freckle number. Other genes and the environment influence freckle size, color, and pattern. For example, sun exposure can temporarily cause more freckles to appear.

Curly Hair

Round hair follicles make straight hair, flattened or c-shaped hair follicles make curly hair, and oval hair follicles make wavy hair. Hair texture is a continuous trait, meaning that hair can be straight or curly or anywhere in between.

Curly hair is influenced by genes much more than by the environment. While curly hair runs in families—people with curly hair tend to have children with curly hair—its inheritance patterns are often unpredictable.

Multiple genes control hair texture, and different variations in these genes are found in different populations. For instance, curly hair is common in African populations, rare in Asian populations, and in-between in Europeans. Straight hair in Asians is mostly caused by variations in two genes—different genes from the ones that influence hair texture in Europeans. And different genetic variations make hair curly in African and European populations.

Hand Clasping

Without thinking about it, fold your hands together by interlocking your fingers. Which thumb is on top—your left or your right?

One study found that 55% of people place their left thumb on top, 45% place their right thumb on top, and 1% have no preference. A study of identical twins concluded that hand clasping has a strong genetic basis (most twins share the trait), but it doesn't fit a predictable inheritance pattern. It is likely affected by multiple genes as well as environmental factors.

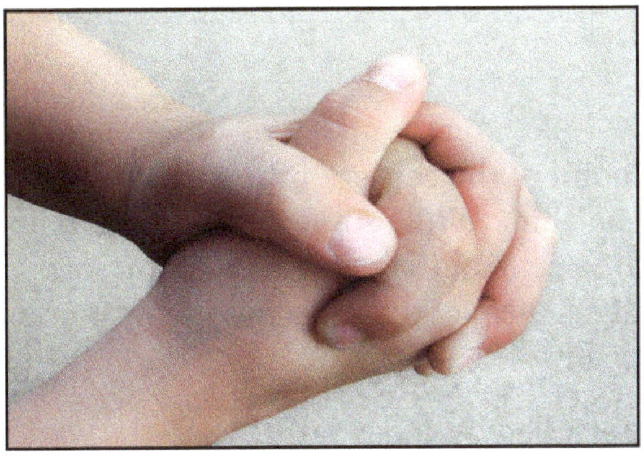

Red/Green Colorblindness

Red-green colorblindness is caused by a single gene located on the X-chromosome. This gene codes for a protein in the eye that detects certain colors of light. When this gene is defective, the eye cannot differentiate between red and green.

You need at least one working copy of the gene to be able to see red and green. Since boys have just one X-chromosome, which they receive from their mother, inheriting one defective copy of the gene will render them colorblind. Girls have two X-chromosomes; to be colorblind they must inherit two defective copies, one from each parent. Consequently, red-green colorblindness is much more frequent in boys (1 in 12) than in girls (1 in 250).

Red-green color blindness follows a very predictable recessive, sex-linked inheritance pattern. A woman with one defective copy of the gene and one functional copy, even though she is not colorblind herself, is known as a "carrier." She has a 50% chance of passing the defective copy to each of her children. Half of her sons will be colorblind, and half of her daughters will be carriers.

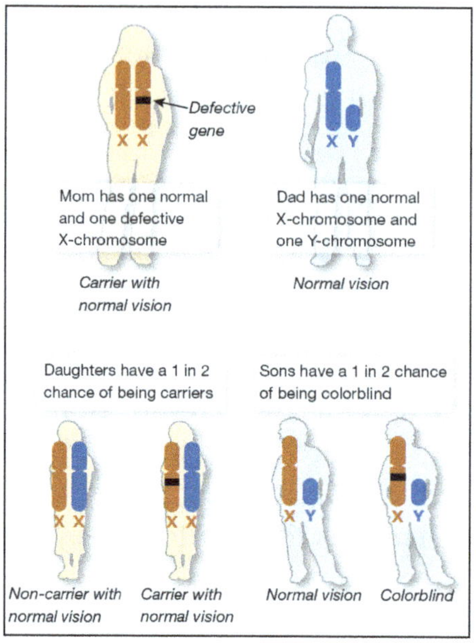

Hairline Shape

If your hairline forms a point at the center of the forehead, you have a widow's peak. If not, you have a straight hairline. While some sources say that widow's peak is a dominant trait controlled by one gene, no scientific study supports this claim. Complicating the question of heritability is the fact that the trait is continuous: some people have just a slight suggestion of a peak.

Widow's peak is likely controlled by genes rather than the environment. But while hairline shape tends to run in families, its pattern of inheritance is usually unpredictable, suggesting that multiple genes are involved.

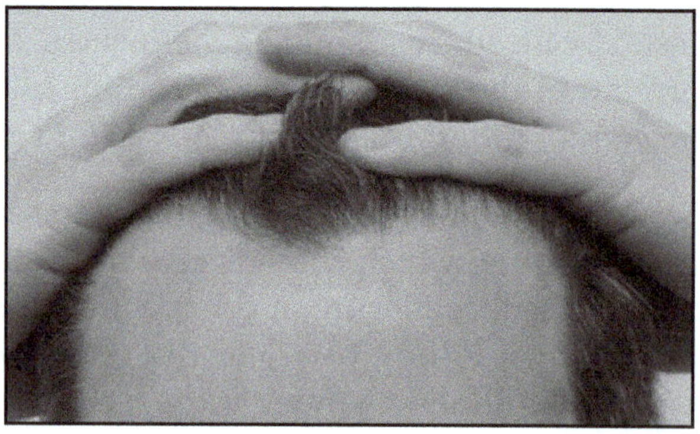

PTC Tasting

To about 75% of us, the chemical PTC (phenylthiocarbamide) tastes very bitter. For the other 25%, it is tasteless. The ability to taste PTC is controlled mainly by a single gene that codes for a bitter-taste receptor on the tongue. Different variations, or alleles, of this gene control whether PTC tastes bitter or not.

PTC tasting follows a very predictable pattern of inheritance. Tasting is dominant, meaning that if you have at least one copy of the tasting version of the gene, you can taste PTC. Non-tasters have two copies of the non-tasting allele.

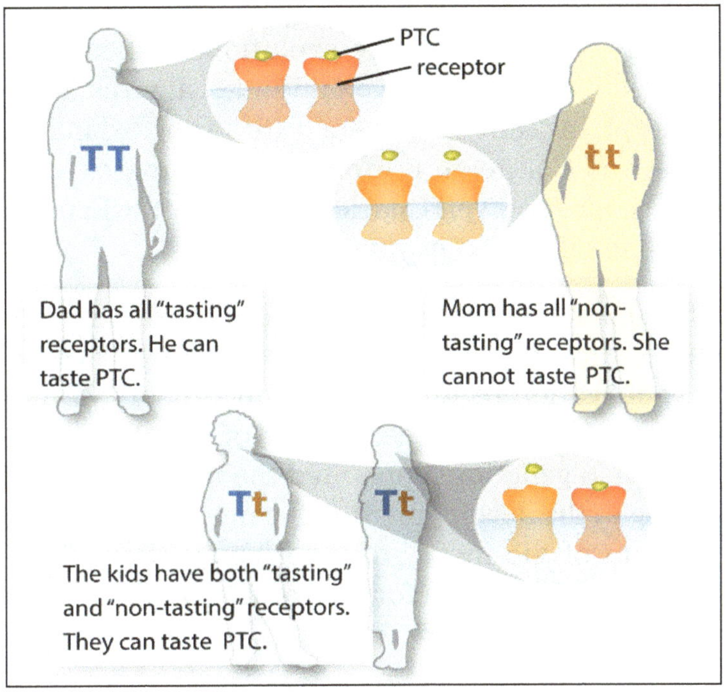

Acquired Traits

Jean-Baptiste Lamarck.

Inheritance of acquired characteristics or inheritance of acquired characters is the once widely accepted idea that physiological modifications acquired by an organism can be inherited by the offspring. Acquired characteristics (or characters) are those changes in the structure or function of an organism that are the result of use, disuse, environmental influences, disease, mutilation, and so forth, such as a muscle that is enlarged through use or mice that have their tails cut off. The theory of the inheritance of acquired characteristics, or "soft inheritance," holds that an organism experiencing such a modification can transmit such a character to its offspring.

This theory is commonly equated with the evolutionary theory of French naturalist Jean-Baptiste Lamarck known as Lamarckism. While Lamarck is sometimes viewed as the founder of the concept, in reality this theory was proposed in ancient times by Hippocrates and Aristotle, and Comte de Buffon, before Lamarck, proposed ideas about evolution involving the concept. Even Charles Darwin, after Lamarck, discussed positively this view in his seminal work, *Origin of Species*.

While the theory of the inheritance of acquired characteristics was enormously popular during the early nineteenth century as an explanation for the complexity observed in living systems, after publication of Charles Darwin's theory of natural selection, the importance of individual efforts in the generation of adaptation was considerably diminished. Later, Mendelian genetics supplanted the notion of inheritance of acquired traits, eventually leading to the development of the modern evolutionary synthesis, and the general abandonment of the theory of inheritance of acquired characteristics in biology, although there are proponents for its working on the microbial level or in epigenetic inheritance.

However, in a wider context, the theory of inheritance of acquired characteristics does remain a useful concept when examining the evolution of cultures and ideas and is generally held in terms of some religious views, such as karma and inheritance of sin. In the case of religion, many traditions accept the view that there is an inheritance of acquired spiritual traits—that actions taken in one's life can be passed down in the form of spiritual merit or demerit to one's lineage.

Lamarckism and the inheritance of acquired characteristics

Lamarckism or Lamarckian evolution is a theory put forward by Lamarck based on the heritability of acquired characteristics. Lamarck proposed that individual efforts during the lifetime of the organisms were the main mechanism driving species to adaptation, as they supposedly would acquire adaptive changes and pass them on to offspring.

The identification of "Lamarckism" with the inheritance of acquired characteristics is regarded by some as an artifact of the subsequent history of evolutionary thought, repeated in textbooks without analysis. Stephen Jay Gould wrote that late nineteenth century evolutionists "re-read Lamarck, cast aside the guts of it and elevated one aspect of the mechanics—inheritance of acquired characters—to a central focus it never had for Lamarck himself". He argued that "the restriction of "Lamarckism" to this relatively small and non-distinctive corner of Lamarck's thought must be labeled as more than a misnomer, and truly a discredit to the memory of a man and his much more comprehensive system". Gould advocated defining "Lamarckism" more broadly, in line with Lamarck's overall evolutionary theory.

Lamarck's Theory

The evolution of giraffe necks is often used as the example in explanations of Lamarckism.

Between 1794 and 1796 Erasmus Darwin, grandfather of Charles Darwin, wrote *Zoönomia* suggesting "that all warm-blooded animals have arisen from one living filament," and "with the power of acquiring new parts" in response to stimuli, with each round of "improvements" being inherited by successive generations.

Subsequently Lamarck proposed in his *Philosophie Zoologique* of 1809 the theory that characteristics that were "needed" were acquired (or diminished) during the lifetime of an organism were then passed on to the offspring. He saw this resulting in the development of species in a progressive chain of development towards higher forms.

Lamarck based his theory on two observations, in his day considered to be generally true:

1. Individuals lose characteristics they do not require (or use) and develop characteristics that are useful.
2. Individuals inherit the traits of their ancestors.

With this in mind, Lamarck developed two laws:

1. Law of use and disuse: "In every animal which has not passed the limit of its development, a more frequent and continuous use of any organ gradually strengthens, develops and enlarges that organ, and gives it a power proportional to the length of time it has been so used; while the permanent disuse of any organ imperceptibly weakens and deteriorates it, and progressively diminishes its functional capacity, until it finally disappears."
2. Inheritance of acquired traits: "All the acquisitions or losses wrought by nature on individuals, through the influence of the environment in which their race

> has long been placed, and hence through the influence of the predominant use or permanent disuse of any organ; all these are preserved by reproduction to the new individuals which arise, provided that the acquired modifications are common to both sexes, or at least to the individuals which produce the young."

Examples of Lamarckism would include:

- Giraffes stretching their necks to reach leaves high in trees strengthen and gradually lengthen their necks. These giraffes have offspring with slightly longer necks (also known as "soft inheritance").
- A blacksmith, through his work, strengthens the muscles in his arms. His sons will have similar muscular development when they mature.

In essence, a change in the environment brings about change in "needs" *(besoins),* resulting in change in behavior, bringing change in organ usage and development, bringing change in form over time—and thus the gradual transmutation of the species. While such a theory might explain the observed diversity of species and the first law is generally true, the main argument against Lamarckism is that experiments simply do not support the second law—purely "acquired traits" do not appear in any meaningful sense to be inherited. For example, a human child must learn how to catch a ball even though his or her parents learned the same feat when they were children.

The argument that instinct in animals is evidence for hereditary knowledge is generally regarded within science as false. Such behaviors are more probably passed on through a mechanism called the Baldwin effect. Lamarck's theories gained initial acceptance because the mechanisms of inheritance were not elucidated until later in the nineteenth century, after Lamarck's death.

Several historians have argued that Lamarck's name is linked somewhat unfairly to the theory that has come to bear his name, and that Lamarck deserves credit for being an influential early proponent of the *concept* of biological evolution, far more than for the *mechanism* of evolution, in which he simply followed the accepted wisdom of his time. Lamarck died 30 years before the first publication of Charles Darwin's *Origin of Species*. As science historian Stephen Jay Gould has noted, if Lamarck had been aware of Darwin's proposed mechanism of natural selection, there is no reason to assume he would not have accepted it as a more likely alternative to his "own" mechanism. Note also that Darwin, like Lamarck, lacked a plausible alternative mechanism of inheritance—the particulate nature of inheritance was only to be observed by Gregor Mendel somewhat later, published in 1866. Its importance, although Darwin cited Mendel's paper, was not recognized until the modern evolutionary synthesis in the early 1900s. An important point in its favor at the time was that Lamarck's theory contained a mechanism describing how variation is maintained, which Darwin's own theory lacked.

Proponents

Lamarck founded a school of French *Transformationism* which included Étienne Geoffroy Saint-Hilaire, and which corresponded with a radical British school of comparative anatomy based at the University of Edinburgh, which included the surgeon Robert Knox and the anatomist Robert Edmund Grant. Professor Robert Jameson wrote an anonymous paper in 1826 praising "Mr. Lamarck" for explaining how the higher animals had "evolved" from the "simplest worms"—this was the first use of the word "evolved" in a modern sense. As a young student Charles Darwin was tutored by Grant, and worked with him on marine creatures.

The *Vestiges of the Natural History of Creation,* authored by Robert Chambers and published anonymously in England in 1844, proposed a theory modeled after Lamarckism, causing political controversy for its radicalism and unorthodoxy, but exciting popular interest and paving the way for Darwin.

Darwin's *Origin of Species* proposed natural selection as the main mechanism for development of species, but did not rule out a variant of Lamarckism as a supplementary mechanism. Darwin called his Lamarckian hypothesis Pangenesis, and explained it in the final chapter of his book *Variation in Plants and Animals under Domestication,* after describing numerous examples to demonstrate what he considered to be the inheritance of acquired characteristics. Pangenesis, which he emphasized was a hypothesis, was based on the idea that somatic cells would, in response to environmental stimulation (use and disuse), throw off 'gemmules' which traveled around the body (though not in necessarily in the bloodstream). These pangenes were microscopic particles that supposedly contained information about the characteristics of their parent cell, and Darwin believed that they eventually accumulated in the germ cells where they could pass on to the next generation the newly acquired characteristics of the parents.

Darwin's half-cousin, Francis Galton carried out experiments on rabbits, with Darwin's cooperation, in which he transfused the blood of one variety of rabbit into another variety in the expectation that its offspring would show some characteristics of the first. They did not, and Galton declared that he had disproved Darwin's hypothesis of Pangenesis, but Darwin objected, in a letter to "Nature" that he had done nothing of the sort, since he had never mentioned blood in his writings. He pointed out that he regarded pangenesis as occurring in Protozoa and plants, which have no blood (Darwin 1871). With the development of the modern synthesis of the theory of evolution and a lack of evidence for either a mechanism or even the heritability acquired characteristics, Lamarckism largely fell from favor.

In the 1920s, experiments by Paul Kammerer on amphibians, particularly the midwife toad, appeared to find evidence supporting Lamarckism, but were discredited as having been falsified. In *The Case of the Midwife Toad,* Arthur Koestler surmised that the specimens had been faked by a Nazi sympathizer to discredit Kammerer for his political views.

A form of "Lamarckism" was revived in the Soviet Union of the 1930s when Trofim Lysenko promoted Lysenkoism which suited the ideological opposition of Joseph Stalin to Genetics. This ideologically driven research influenced Soviet agricultural policy which in turn was later blamed for crop failures.

Since 1988 certain scientists have produced work proposing that Lamarckism could apply to single celled organisms. The discredited belief that Lamarckism holds for higher order animals is still clung to in certain branches of new-age pseudoscience under the term racial memory.

Steele produced some indirect evidence for somatic transfer of antibody genes into sex cells via reverse transcription. Homologous DNA sequences from VDJ regions of parent mice were found in germ cells and then their offspring.

Neo-Lamarckism

Unlike neo-Darwinism, the term neo-Lamarckism refers more to a loose grouping of largely heterodoxical theories and mechanisms that emerged after Lamarck's time, than to any coherent body of theoretical work.

In the 1920s, Harvard University researcher William McDougall studied the abilities of rats to correctly solve mazes. He claimed that offspring of rats that had learned the maze were able to run it faster. The first rats would get it wrong an average of 165 times before being able to run it perfectly each time, but after a few generations it was down to 20. McDougall attributed this to some sort of Lamarckian evolutionary process.

At around the same time, Russian physiologist Ivan Pavlov, who was also a Lamarckist, claimed to have observed a similar phenomenon in animals being subject to conditioned reflex experiments. He claimed that with each generation, the animals became easier to condition.

Neither McDougall nor Pavlov suggested a mechanism to explain their observations.

Soma to Germ Line Feedback

In the 1970s, the immunologist Ted Steele, formerly of the University of Wollongong, and colleagues, proposed a neo-Lamarckian mechanism to try and explain why homologous DNA sequences from the VDJ gene regions of parent mice were found in their germ cells and seemed to persist in the offspring for a few generations. The mechanism involved the somatic selection and clonal amplification of newly acquired antibody gene sequences that were generated via somatic hyper-mutation in B-cells. The mRNA products of these somatically novel genes were captured by retroviruses endogenous to the B-cells and were then transported through the blood stream where they could breach the soma-germ barrier and retrofect (reverse transcribe) the newly acquired genes into the cells of the germ line. Although Steele was advocating this theory for the better part of two decades, little more than indirect evidence was ever acquired to support it. An interesting attribute of this idea is that it strongly resembles Darwin's own theory of pangenesis, except in the soma to germ line feedback theory; pangenes are replaced with realistic retroviruses.

Epigenetic Inheritance

Forms of 'soft' or epigenetic inheritance within organisms have been suggested as neo-Lamarckian in nature by such scientists as Eva Jablonka and Marion J. Lamb. In addition to "hard" or genetic inheritance, involving the duplication of genetic material and its segregation during meiosis, there are other hereditary elements that pass into the germ cells also. These include things like methylation patterns in DNA and chromatin marks, both of which regulate the activity of genes. These are considered "Lamarckian" in the sense that they are responsive to environmental stimuli and can differentially effect gene expression adaptively, with phenotypic results that can persist for many generations in certain organisms. Although the reality of epigenetic inheritance is not doubted (as countless experiments have validated it) its significance to the evolutionary process is however uncertain. Most neo-Darwinians consider epigenetic inheritance mechanisms to be little more than a specialized form of phenotypic plasticity, with no potential to introduce evolutionary novelty into a species lineage.

Lamarckism and Single-celled Organisms

While Lamarckism has been discredited as an evolutionary influence for larger life forms, some scientists controversially argue that it can be observed among microorganisms. Whether such mutations are directed or not also remains a point of contention.

In 1988, John Cairns at the Radcliffe Infirmary in Oxford, England, and a group of other scientists renewed the Lamarckian controversy (which by then had been a dead debate for many years). The group took a mutated strain of *E. coli* that was unable to consume the sugar lactose and placed it in an environment where lactose was the only food source. They observed over time that mutations occurred within the colony at a rate that suggested the bacteria were overcoming their handicap by altering their own genes. Cairns, among others, dubbed the process adaptive mutagenesis.

If bacteria that had overcome their own inability to consume lactose passed on this "learned" trait to future generations, it could be argued as a form of Lamarckism; though Cairns later chose to distance himself from such a position. More typically, it might be viewed as a form of ontogenic evolution.

There has been some research into Lamarckism and prions. A group of researchers, for example, discovered that in yeast cells containing a specific prion protein Sup35, the yeast were able to gain new genetic material, some of which gave them new abilities such as resistance to a particular herbicide. When the researchers mated the yeast cells with cells not containing the prion, the trait reappeared in some of the resulting offspring, indicating that some information indeed was passed down, though whether or not the information is genetic is debatable: trace prion amounts in the cells may be passed to their offspring, giving the appearance of a new genetic trait where there is none.

Finally, there is growing evidence that cells can activate low-fidelity DNA polymerases in times of stress to induce mutations. While this does not directly confer advantage to the organism on the organismal level, it makes sense at the gene-evolution level. While the acquisition of new genetic traits is random, and selection remains Darwinian, the active process of identifying the necessity to mutate is considered to be Lamarckian.

Sexual Selection

Sexual selection is a "special case" of natural selection. Sexual selection acts on an organism's ability to obtain (often by any means necessary) or successfully copulate with a mate.

Selection makes many organisms go to extreme lengths for sex: peacocks (top left) maintain elaborate tails, elephant seals (top right) fight over territories, fruit flies perform dances, and some species deliver persuasive gifts. After all, what female Mormon cricket (bottom right) could resist the gift of a juicy sperm-packet? Going to even more extreme lengths, the male redback spider (bottom left) literally flings itself into the jaws of death in order to mate successfully.

Sexual selection is often powerful enough to produce features that are harmful to the individual's survival. For example, extravagant and colorful tail feathers or fins are likely to attract predators as well as interested members of the opposite sex.

Charles Darwin proposed that all living species were derived from common ancestors. The primary mechanism he proposed to explain this fact was natural selection: that is, that organisms better adapted to their environment would benefit from higher rates of survival than those less well equipped to do so. However he noted that there were many examples of elaborate, and apparently non-adaptive, sexual traits that would clearly not aid in the survival of their bearers. He suggested that such traits might evolve if they are sexually selected, that is if they increase the individual's reproductive success, even at the expense of their survival.

Darwin noted that sexual selection depends on the struggle between males to access females. He recognized two mechanisms of sexual selection: intrasexual selection, or competition between members of the same sex (usually males) for access to mates, and intersexual selection, where members of one sex (usually females) choose members of the opposite sex. The idea of cumbersome traits evolving to aid males in competition during aggressive encounters was readily accepted by scientists shortly after Darwin's publication. However, the idea of female mate choice was received with ridicule, and was not seriously reconsidered until nearly 80 years later. In the 40 years since, there has been much progress in our understanding of how sexual selection operates.

Sex that is under Stronger Selection

Sex roles are defined by differences in gametes: females produce relatively few, highly nutritious (usually non-motile) gametes, whereas males produce comparatively abundant, smaller, motile gametes. Because only a single gamete of each type is required to produce an offspring, there will be an excess of male gametes that will not fertilize any eggs. This asymmetry leads to Bateman's principle, whereby female reproduction is primarily limited by their access to resources to nourish and produce these large gametes, whereas male reproduction is mainly limited by access to females. Therefore males typically compete among themselves for access to females, whereas females tend to be choosy and mate only with preferred males.

In sexually reproducing species, every offspring has one father and one mother, so the average reproductive success is equal for both males and females. A successful male can potentially sire many offspring. If a male gains a disproportionate share of reproduction, he will take away reproductive opportunities from other males, leading to a high reproductive variance among males. A successful female, on the other hand, will not take away reproductive opportunities from other females, leading to a smaller variance in reproductive success. The higher the reproductive variance, the stronger the effects of sexual selection. Strong sexual selection typically results in sexually dimorphic traits that are exaggerated, or more elaborate, in the sex with highest reproductive variance.

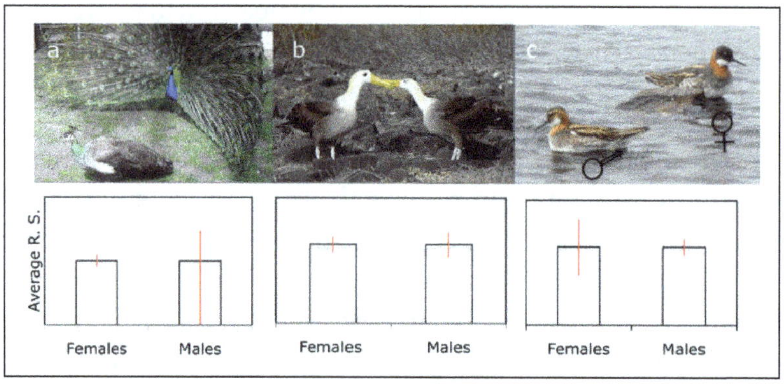

Figure: Variance in reproductive success explains which sex is subject to stronger sexual selection.

Males and females in a population have the same average reproductive success (R. S., black bars) but they may differ in the reproductive variance among members of each sex (shown in red). Differences in the selection gradient will result in sexual dimorphism.

(A) When males are subject to stronger sexual selection than females, males will evolve secondary sexual characters that result in marked differences between the sexes. Peacocks do not provide any parental care, and some males are more successful than others who may never reproduce, leading to marked dimorphism.

(B) When males contribute to offspring care, the selection gradient is lower and the sexes will be monomorphic. Many seabirds are monogamous and raise offspring together and the sexes are indistinguishable.

(C) When males provide all the parental care, the selection gradient can be reversed and females may have to compete for access to males, leading to reverse sexual dimorphism. Red-necked phalaropes compete for access to males who provide all the parental care. Females are larger and more aggressive than males.

The Role of Parental Care

Most species provide little or no care to their offspring, but in species where parental care is required, variance in reproductive success will be impacted not only by fertilization success, but also by the contribution of each sex to the care of the offspring. The degree and direction of sexual dimorphism can be explained by the relative selection gradients of each sex. If females provide more parental care than males, the variance in male reproductive success can be expected to be large, since females providing offspring care will not be immediately available for further reproduction and competition for available females will increase among males. The exaggerated tail of the (male) peacock compared to the shorter tail of the (female) peahen, indicates that males are under stronger sexual selection than females. However, in species where biparental care is required to successfully raise offspring, variance in male reproductive success is generally lower, since males that are engaged in providing parental care will not be able to invest as much energy in pursuing additional mating opportunities. This situation often results in the emergence of sexually monomorphic species, in which the male and female look and behave in similar ways. Finally, in rare situations where only males provide parental care, males can become the limiting resource for females. Under these circumstances, the variance in reproductive success may be high for females who then tend to monopolize access to one or more males to care for their offspring. Males may then become choosy about which females they mate with. This often results in reversed sexual dimorphism, such that females evolve more elaborate secondary sexual characters than males.

Mechanism of Sexual Selection Operate

Sexual selection can operate both intra- and inter-sexually, either sequentially or simultaneously. During intrasexual selection, members of the same sex attempt to outcompete rivals, often during direct encounters. Intrasexual selection is typically responsible for the evolution of male armaments such as deer antlers, beetle horns, and large body size that provide individuals with an advantage when fighting off potential competitors. Individuals, who are better able to exclude competitors, have a greater chance to acquire mates and father offspring. For example, dominant male red deer monopolize a group of females (also known as harem) by constantly fighting off competitors, and they father most of the offspring produced by the females. By contrast, intersexual selection results from interactions between the sexes, typically involving mate choice. The evolution of elaborate behavioral displays and morphological traits can often be explained as the result of intersexual selection. Usually, females tend to be more choosy, evaluating morphological and behavioral traits from potential mates to determine which will maximize their fitness. Males tend to compete with one another to gain the female's attention. An extreme example of intersexual selection can be found in species where males form leks where multiple males gather to display to females.

Sexual selection episodes can occur before mating takes place (pre-copulatory), or during and after mating (post-copulatory), and they can occur within a sex (intrasexual) and between the sexes (intersexual).

	Prc-cooulato""	Post conulato...,
lntrascxuat SClcct on (competition)	• To monopolize a harem: Elephant seals (Le Boeuf 1974) • To monopolize a sing e fema e: Rhinoceros Beetles (Emlen 2008)	• Greater relative testis size in species with greater risk of female mutiple mating (Calh m and Blrkheod 2007)
Interscxual Selectlon (mate choice)	• Males displaying at leks in Manakins (Prum 1990)	• Female reproductive morphology that prevents eja-culated sperm from contacting eggs directly (Keler and Reeve 1995)
SexualConflict	• Intimidation and sexual harassment (Clutton-Brock and Parker 1995)	• Traumatic insemination in bed bugs (Stutt and Siva-jothy 2001)

Table: Some examples of when and how sexual selection operates.

Choosing a Mate

Why do females choose between males rather than mate at random, or with the first male they encounter? Females can directly increase their reproductive success by mating with certain, select males and acquiring direct benefits. For example, females can gain increased access to food, protection from harassing males, or help in raising offspring, and avoid being infected with parasites or other diseases by choosing healthy males. However there are instances where females do not appear to gain any direct benefit from males, yet they still discriminate among them. Under these conditions, females likely gain indirect benefits via their offspring. These indirect benefits are usually genetic rather than resource based. By choosing certain males, their offspring will likely inherit genes that tend to increase their fitness. Males often evolve traits and displays that advertise their ability to provide direct and indirect benefits, and females evolve preferences for these traits. Two major mechanisms to account for female mate choice have been proposed: good genes, and Fisherian arbitrary processes.

Good Genes

Under the 'good genes' scenario, differences among males provide females with information about the genetic qualities of the different males that can be inherited by the offspring. Under the 'good genes,' just as in the 'direct benefits' models, there is correspondence between the putative roles of natural versus sexual selection, since preferring certain males can result in a female gaining higher viability, fecundity, and reproductive success, for her offspring. Good genes can be those that allow males to carry a 'handicap,' yet survive despite having a cumbersome trait, genes that signal resistance to diseases, or genes that are more compatible with those of the female. Evidence of female choice for good genes remains scarce despite decades of studies of female mate choice in many taxa. This apparent lack of success continues to create debate as to the importance of the good genes model in the field.

Fisherian Arbitrary Choice

Named after R. A. Fisher, who originally proposed it, this model suggests that female preference can evolve for arbitrary traits that do not provide information about the male's quality, and that therefore do no reinforce the effects of natural selection. If females evolve a preference for a particular trait, males bearing that trait will be selected as mates. This assortative mating will establish a genetic correlation between the preference and the trait. The fitness advantage of the arbitrary trait exists only as a result of its covariance with the preference. By choosing a male with a particular trait value, the females gain the indirect benefit of producing offspring that will be more sexually attractive to females that carry the preference. This type of process can result in a runaway positive feedback loop, whereby the trait becomes more exaggerated as selection on the preference increases, but other models have shown that such a feedback loop is only one of many possible evolutionary outcomes of the Fisherian process.

Instances when Sexual Selection Act

Sexual selection can affect reproductive success at multiple reproductive stages. First, it acts during all the processes that lead to acquiring mating opportunities (i.e., excluding competitors, attracting, selecting and retaining mates). Darwin referred exclusively to pre-copulatory sexual selection in his discussions, erroneously assuming that mating would inevitably result in reproductive success. In recent years, evidence that copulatory and post-copulatory events play an important role in determining the outcome of fertilization and reproduction has been increasing. Post-copulatory selection refers to the events that occur during and after mating. Post-copulatory male-male competition is known as sperm competition a term coined by Parker who recognized that when females mate with multiple males, their ejaculates compete inside the female reproductive tract for access to eggs. Sperm competition has resulted in the evolution of morphologically modified sperm that increase the likelihood of fertilization in many taxa. Post-copulatory female choice refers to the ability of females to affect the likelihood that sperm from a particular male fertilizes their eggs, and their decision to invest in offspring based on the identity of the male with whom they mate. Females exert this choice via morphological, chemical and behavioral adaptations. This type of selection is called cryptic choice because it occurs inside the female reproductive tract and cannot be detected from behavioral studies alone.

Conflict between the Sexes

Although both sexes are seeking to optimize their reproductive success, their genetic interests are not aligned, resulting in sexual conflict. Traits that allow a male to increase his reproductive success at the expense of the female will be positively selected if the female mates with multiple males. These traits will be genetically transmitted and spread in the population, despite their negative effects on female reproductive success, if the reproductive success of these males is higher than that of males lacking such traits. Sexual conflict can often result in an evolutionary arms race, whereby the evolution of a trait that imposes harm on one sex will result in evolution of a counter-trait to mitigate the harm on the affected sex, with subsequent escalation in both. Examples of sexual conflict include traumatic insemination in bed bugs, copulatory grasping and anti-grasping structures in waterstriders, and genital coevolution in waterfall.

Stabilizing Selection

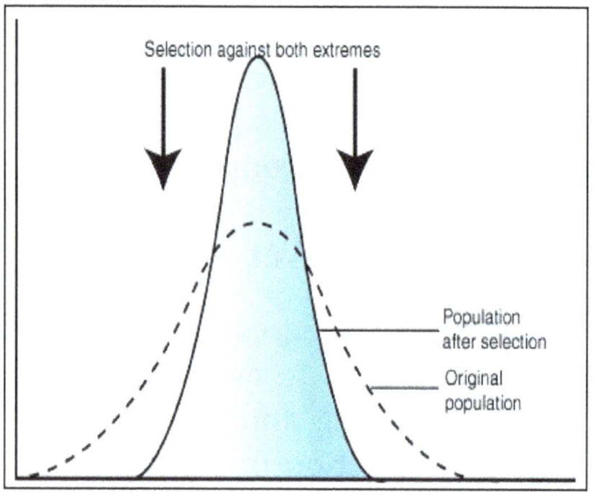

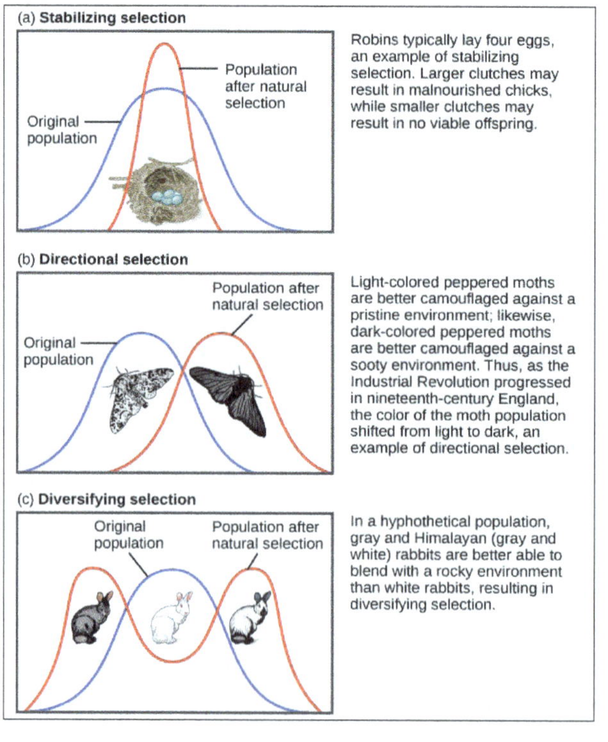

Stabilizing Selection Examples

Stabilizing selection is any selective force or forces which push a population toward the average, or median trait. Stabilizing selection is a descriptive term for what happens to an individual trait when the extremes of the trait are selected against. This increases the frequency of the trait in the population, and the alleles and genes which help form it. Many traits which are common across entire groups of species have been formed through the effects of stabilizing selection. Stabilizing selection can be seen in the image below, comparing the three types of selection.

Robin Eggs

In this case, the number of eggs in a robin nest has been selected for through a stabilizing selection. Robins are apparently not ably to raise more than 4 chicks with much success. This is probably because of the size of the birds and the amount of food that two adults can provide. By comparison, most penguins can only raise one chick at a time, due to the size of the chick and the amount of food it requires. While they stabilized on different numbers, both are forms of stabilizing selection which maximize the fitness of the species in their environment.

As opposed to the other forms of selection, you can clearly see in the stabilizing selection graph that the population of the median trait increases, while the other populations decrease. In this case, 5 eggs is too many and some would die. On the other hand, 3 is too few. Either the eggs are not viable enough to only rely on 3 eggs, or predation and other forces require more than 3 eggs to overcome them and carry on to another generation.

Hypothetical Lemurs

There is a population of multicolored lemurs on Crazy Island. This particular population of lemurs has been observed by scientists, and they have noticed the following changes in the lemur's color.

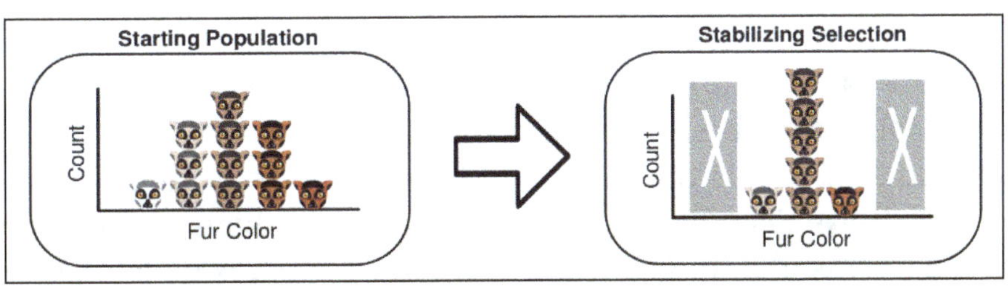

As you can see, the lemurs have obviously undergone a stabilizing selection. The light and dark lemurs have almost disappeared, while the middle brown lemurs have increased. Without further information, it is not clear why this would be the case. It is the job of ecologists and evolutionary biologists to observe the population, noting peculiar aspects of the various forms to understand what may have caused the stabilizing selection. This is no easy question to answer and may have many more than one answer.

In the case if the lemurs, it could be that the darker and lighter lemurs were both easier to spot by predators. If the lemurs only have one predator, this is as easy hypothesis to test. A scientist would simply observe the predator, and see which lemurs it prefers. This would lend evidence to the hypothesis that the stabilizing selection is caused by *predation*. Further evidence could include the amount of lemurs the predators eat, and models showing how that level of predation could produce the coloration seen.

However, it is much more common that a species has multiple selective pressures, and that each pressure acts on various traits in different ways. For example, the lighter color could be suffering from predation, whereas the darker version could be overheating. (Dark colors absorb more solar heat). Likewise, predation could be driving both traits, but not influencing them totally. Female lemurs might have a preference for brown lemurs, due to their increased survival. This would be a form of *sexual selection*, driving a trend of stabilizing selection.

Common Causes of Stabilizing Selection

Stabilizing selection, along with *directional selection* and *disruptive selection*, refer to the direction of individual traits. While stabilizing selection pushed the trait towards the average instead of one or both of the extremes, it can be driven by any form of selection. Some of the most common forms of selection are from predation, resource allocation, coloration of the environment, food type, and a wide variety of other forces.

Many traits we don't talk about regularly have been driven by a variety of causes throughout history. Take the modest insect, for example. All insects have an *exoskeleton*, a miraculous structure made of *chitin* and other structural molecules which form a shield around their organs and allow them to maintain a water balance in the harshest of environments. This shield, while it has been modified in to an almost infinite number of forms, was first selected for out of stabilizing selection. The ancestors to insects did not have this *adaptation*, but once it evolved it was highly favored.

Simply stated, there is no common cause of stabilizing selection, besides the fact that the most average individual is selected for. In that way, like all forms of selection, the cause of stabilizing selection is the increased fitness and reproductive success that the median individuals have. The extreme versions or traits have a disadvantage, in one way or another. This disadvantage, in evolutionary terms, is decreased reproduction. The traits they carry are coded for in part by their DNA, which they can only pass on through reproduction. In stabilizing selection, the increase in the median traits represents their increased success. The other extreme traits are not as successful, possibly causing their owners to die. This increases the resources available to the median animals, further boosting their success. In this way, stabilizing selection is the cause of many traits that entire groups of animals share. These are known as *synapomophies*.

Directional Selection

Directional selection occurs when individuals with traits on one side of the mean in their population survive better or reproduce more than those on the other. It has been demonstrated many times in natural populations, using both observational and experimental approaches. Directional selection does the "heavy lifting" of evolution by tending to move the trait mean toward the optimum for the environment. It results in increased adaptedness of organisms. It is the principle process that Charles Darwin himself envisaged as driving adaptive evolution. Two of Darwin's own examples were:

(1) Faster wolves being more successful at hunting deer,

(2) Flowers that produce more nectar being more successful in attracting pollinating insects.

These both suggest directional rather than other forms ("modes") of selection. Directional selection is the process that comes most easily to mind when thinking about natural selection, and it is the form of selection that has taken place in the best-known examples of evolution (e.g., the peppered moth, antibiotic resistance, finch beaks). However, directional selection does not always result in evolution, because it can be constrained in many ways. If directional selection acts in different directions in different populations or species, because of variation in environmental circumstances, then it is described as *divergent*. This results in populations becoming different, and it can contribute to speciation. Directional selection can also be artificially imposed, and it has commonly been used by animal and plant breeders to improve traits (such as yield) in domesticated organisms, as well as to better understand evolution.

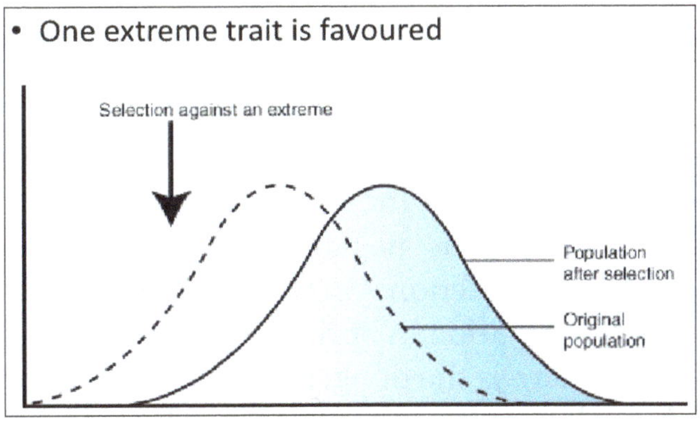

Example: Industrial Melanism.

- In England, before industrialisation, white-winged moth were more in no. than dark-winged moths.
- But after industrialization, dark-winged moths became more in no. than white winged moths.

- This is because during industrialization, the tree trunks covered by white lichen became dark due to deposition of coal & dust particles.
- As a result, white-winged moths can be easily picked up by predators from the dark background & dark-winged moths survived.

Negative Selection

Scientists talk about positive selection when the focus of a particular study is on an increase in rare variants that improve optimal fitness, and they speak of negative selection when the focus is on the removal of harmful variants.

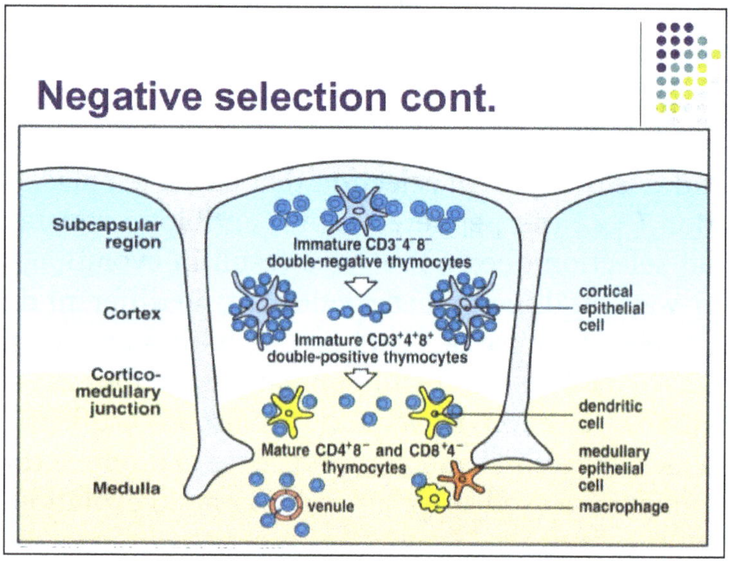

Causes of Negative Selection

Because more DNA changes are harmful than are beneficial, negative selection plays an important role in maintaining the long-term stability of biological structures by removing deleterious mutations. Thus, negative selection is sometimes also called purifying selection or background selection. One key reason why this form of selection is so prevalent is the success of evolution in optimizing biological structures: As soon as a system has been improved, there is the danger of losing that improvement by a deleterious mutation. Purifying selection makes sure that deleterious mutations cannot take over a population and that any improved structures—once fixed in a population—are maintained as long as they are needed. A dramatic example of such maintenance can be found in so-called "living fossils": If a species' ecological niche happens not to change for millions of years, fossil forms of the species can be almost indistinguishable from their present-day descendants.

More short-term negative selection is also widespread, especially due to ecological reasons. Many structures in biology are only conditionally optimal, because they depend upon the details of other structures or circumstances to perform their function. If such other structures are within the same organism, this relationship is termed epistasis. For example, two proteins could interact epistatically in such a way that a deleterious mutation in one protein could be either compensated for or aggravated by a mutation in the other protein. Frequently, ecological circumstances also play a role in determining mutational effects. For instance, if the niche of a species stays the same, some mutations that would be beneficial in other niches will be under negative selection. If the niche changes, however, some traits that were previously under negative selection may suddenly be beneficial and have a greater fitness than the majority of the previously favored genotypes.

If environmental interactions include other rapidly evolving species, then the pressure to change may never stop, and the evolutionary optimum will always remain some steps ahead. Host-parasite interactions are a famous example of this sort of situation. Here, the host immune system evolves to recognize a special structure on the parasite and allow its removal. This in turn induces negative selection on the current form of the parasite while leading to positive selection of variants that cannot be recognized by the host. Furthermore, if such variant parasites exist, they will increase in frequency and in turn induce negative selection of the current variant of the host, which will lead to the positive selection of hosts that can again recognize the parasite, and so on. Negative (and positive) selection in such a system never rests, which is why one hypothesis describing these systems was named after the Red Queen in the book *Alice in Wonderland,* who famously stated that it takes a great amount of running to stay in the same place.

Figure: Long term stabilizing selection can explain \"living fossils.\"

Amber fossils sometimes conserve ancient insect species that are very similar to their modern relatives. Voltinia dramba is the first butterfly to be described from a 20 million-year-old fossil. One of its several close relatives is found in Mexico today.

Effective Strength

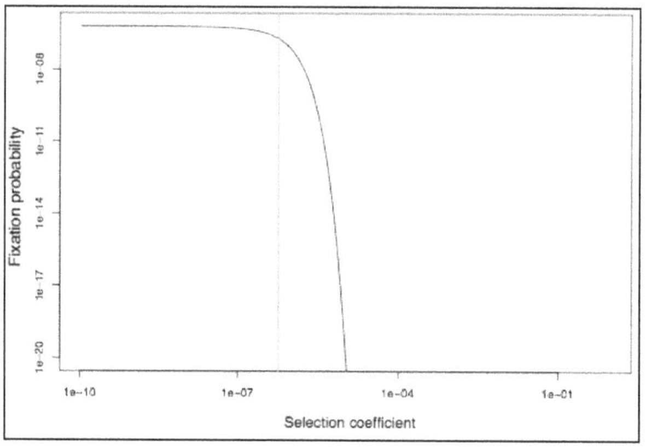

The strength of negative selection.

The fixation probability is given for heterozygous, harmful mutations. The vertical line indicates selection coefficients that satisfy Nes = 0.5 and roughly separates the domains where genetic drift (left) and selection (right) dominate the fate of new mutations. The effective population size Ne was estimated from DNA sequence diversity data in the fruitfly *Drosophila miranda*.

There are two ways to measure the strength of negative selection: an absolute one and a relative one. The absolute selection coefficient *s*quantifies the relative fitness difference between the rare variant (mutant) and the most common variant (wild type). The advantage of this measure is its independence from population size. For mutations with large effects, it might even be possible to measure selection coefficients by comparing direct counts of surviving offspring from many individuals with and without a well-specified mutation.

However, most selection coefficients are very small and, thus, selection as a process is often better understood by looking at the relative effective selection coefficient $N_e s$, which is the product of the absolute selection coefficient and the effective population size N_e. A value of 1 ($N_e s = 1$) approximately denotes a threshold: Mutations with $N_e s < 0.25$ are fixed with probabilities that are comparable to those of neutral mutations, even if they are harmful. On the other hand, mutations in which $N_e s > 4$ will very rarely become fixed in a population with reasonable levels of recombination. Moreover, it is unlikely that a harmful mutation with $N_e s$ exceeding a few dozen could have ever been fixed in the entire history of the known universe. Such deleterious mutations can only be fixed if they are closely linked to a very strongly advantageous mutation that will drag them to high frequencies (but, as discussed later, this reduces N_e and, thus, the rule of thumb remains valid). If harmful mutational effects are smaller than roughly $N_e s = 0.25$, then they can pass below the "radar" of natural selection: Here, natural selection can no longer distinguish among beneficial, neutral, and harmful mutations, so it treats them all in the same way. This problem is balanced by the increasing frequency of advantageous back mutations for such sites. To apply this simple rule of thumb correctly, it is important to note that N_e usually has to be measured from DNA sequence diversity data in the wild. N_e is usually much lower than the number of all existing individuals. Also, this rule has limitations when recombination is rare or absent. In this case, the effective strength of selection is reduced, and more deleterious mutations can fix more easily. Figure shows the fixation probabilities of various deleterious mutations under free recombination.

Consequences of Negative Selection

The main consequence of negative selection is the extinction of less-adapted variants. If the best-adapted variant does not change because it is at a stable local optimum, then negative selection will remove all new variants for that optimal trait.

It is important to note that negative selection can also impact molecular diversity. Consider the simple case of a population with genes that are optimally adapted to the existing constant environment. Such a setting is probably realistic for many "housekeeping" genes that ensure the proper working of the basic molecular machinery of life. Almost every mutation that happens in these genes will be deleterious, and, because mutations are the inevitable consequence of the molecular machinery that copies DNA, we can expect a substantial number of such harmful mutations. The deleterious nature of these mutations will result in their quick removal. In any real-life setting, however, an important side effect of such a removal will be the accompanying removal of linked mutations. This has important implications for the study of molecular diversity, as all neutral mutations linked to deleterious mutations will not be observed in the population and the corresponding N_e values will thus be reduced. This reduction in N_e has consequences for adaptive evolution, because the effective strength of selection for positive mutations will be reduced as well, and more advantageous mutations with small effects will be lost by chance.

If negative selection is too weak to remove harmful mutations, then deleterious mutation accumulation will occur, and a gradual decay of genomic integrity will be the result. This can lead to extinction for some species if it continues long enough; however, the resulting widespread existence of deleterious mutations in such a genome will eventually also lead to the occurrence of back mutations, which (among many other factors) can significantly contribute toward maintenance of a reasonable level of integrity in the genome of other species in the long term.

If negative selection is too strong for the whole population, extinction will occur, unless the population is rescued in time. Extinction can occur if the negative selection considered is "hard" selection, which actually reduces the number of surviving offspring that are produced. "Soft" selection (which occurs when the reproductive capacity of an organism is high enough) can also be negative, but it will lead only to competition over who will increase in frequency within the population, effectively without a reduction of the maximal number of offspring that can be produced. Thus, no extinction risk exists with soft selection.

Climate change and other habitat alterations are currently placing many species under such extreme negative selection that these species' survival is threatened. Also, mutagenic substances that are released into the environment by humans lead to a general increase in the frequency of mutations, a vast majority of which are deleterious and further increase the negative selection pressure for many populations. Thus, understanding the causes, extent, and consequences of negative selection can contribute important insights toward securing biodiversity in the long term.

Gene Selection

Genetic selection is the process by which certain traits become more prevalent in a species than other traits. These traits seen in an organism are due to the genes found on their chromosomes. The genes code for the traits that we are able to observe.

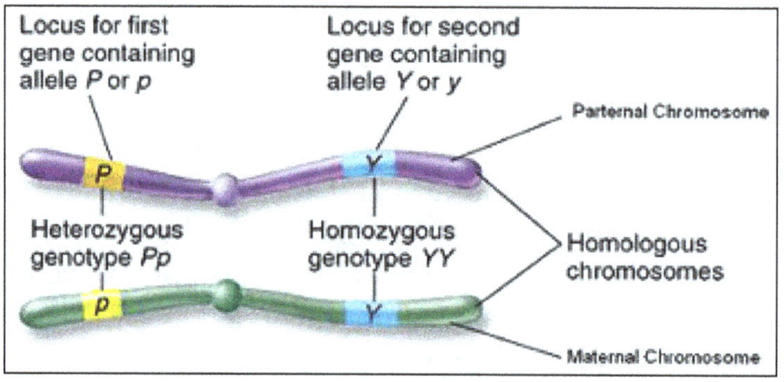

Figure: Alleles for genes are inherited and come in various forms.

Genes have more than one version or allele. We inherit one allele for every gene from each of our parents.

Some alleles are seen more frequently in a population because there are factors that select those genes.

In natural selection, natural forces determine the traits seen in an organism. A variation or allele of a trait makes some individuals more suited to survive in the environment. Mating behavior that leads to a sexual preference for a trait is also natural selection.

If you are a mouse and you can blend in with the environment you are likely to live long enough to pass on your genes for coat color to baby mice. Your friend who could not blend in gets eaten and does not pass on the genes for their coat color. Over time, the mice in dark areas will have dark fur and the mice in light areas will have light fur. Natural selection by the environment eliminates the weakest individuals from the gene pool. If the mice did not have a gene for coat color that allowed them to hide from predators, that species of mouse would be eliminated from the environment.

Natural selection also occurs when a species has a preference for certain traits for sexual reasons. Male peafowl (peacocks) have elaborate tail feathers because female peacocks are attracted to really nice tails. This is called sexual selection since traits are being selected for sexual reasons.

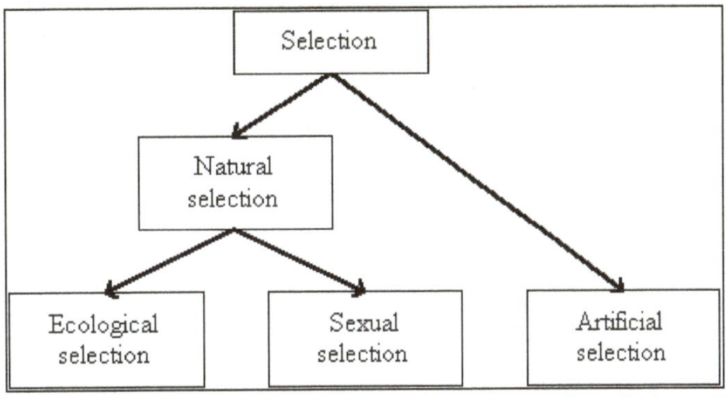

Figure: There are two types of selection: 1) Natural and 2) Artificial.

Gene flow is the movement of genes into or out of a population. Such movement may be due to migration of individual organisms that reproduce in their new populations, or to the movement of gametes (e.g., as a consequence of pollen transfer among plants). In the absence of natural selection and genetic drift, gene flow leads to genetic homogeneity among demes within a metapopulation, such that, for a given locus, allele frequencies will reach equilibrium values equal to the average frequencies across the metapopulation. In contrast, restricted gene flow promotes population divergence via selection and drift, which, if persistent, can lead to speciation.

Gene Selection Theory

Two-toed sloths.

In modern evolutionary theory, all evolutionary processes boil down to an organism's genes. Genes are the basic "units of heredity," or the information that is passed along in DNA that tells the cells and molecules how to "build" the organism and how that organism should behave. Genes that are better able to encourage the organism to reproduce, and thus replicate themselves in the organism's offspring, have an advantage over competing genes that are less able. For example, take female sloths: In order to attract a mate, they will scream as loudly as they can, to let potential mates know where they are in the thick jungle. Now, consider two types of genes in female sloths: one gene that allows them to scream extremely loudly, and another that only allows them to scream moderately loudly. In this case, the sloth with the gene that allows her to shout louder will attract more mates—increasing reproductive success—which ensures that her genes are more readily passed on than those of the quieter sloth.

Essentially, genes can boost their own replicative success in two basic ways. First, they can influence the odds for survival and reproduction of the organism they are in (individual reproductive success or fitness—as in the example with the sloths). Second, genes can also influence the organism to help other organisms who also likely contain those genes—known as "genetic relatives"—to survive and reproduce (which is called inclusive fitness). For example, why do human parents tend to help their own kids with the financial burdens of a college education and not the kids next door? Well, having a college education increases one's attractiveness to other mates, which increases one's likelihood for reproducing and passing on genes. And because parents' genes are in their own children (and not the neighborhood children), funding their children's educations increases the likelihood that the parents' genes will be passed on.

Disruptive Selection

Disruptive selection is an evolutionary force that drives a population apart. The disruptive selection will cause organisms with intermediate traits to reproduce less, and will allow those organisms with extreme traits to reproduce more. This causes the alleles for the extreme traits to increase in frequency. Over time, and with enough disruptive selection, a population can be completely divided. When this happens, the two populations can become diverse enough to form separate species. In the most basic sense, disruptive selection can act on a single gene, selecting between the different alleles present in a population. On a much broader level, disruptive selection can affect a variety of traits and drive a population to become reproductively isolated from the original population.

Disruptive selection, also called *diversifying selection*, is based on the *variance* of a trait in a population. A gene with only one allele would have no variance, and selection could not act on differences in the trait created by the gene. Most genes have many different alleles, which create a wide variety of functions. The result of these many alleles acting in a population and the other genes in affect lead to traits that have a distribution of different types, sizes, or patterns. If the trait has an almost infinite variety of different forms, it is *continuous*. If the trait exists in distinct entities, it is *discrete*. A continuous trait would be height, while a discrete trait might be eye color. Either way, most traits have a high level of variance, due to the interactions of various genes and alleles. Disruptive selection acts on the traits in the middle of the spectrum.

Disruptive selection is usually seen in high-density populations. In these populations resources become scarcer, and competition for the resources increases. This *intraspecific competition* can cause differences between organisms to have a more profound effect on each organism's survival. Selective pressures that might not have factored into a low-density population can take effect, and the resulting disruptive selection can drive a population apart. In doing so, the populations are often pushed to different niches, lowering the competition between them. This leads to *sympatric speciation*, or speciation that occurs while populations occupy the same area.

Examples of Disruptive Selection

Finches on Santa Cruz Island

Darwin's finches, or Galapagos finches, are a group of finches that inhabit the long chain of islands known as the Galapagos, famously visited by Charles Darwin. The birds have been rigorously studied, and various patterns of evolution have been seen in different populations on different islands. On Santa Cruz Island, disruptive selection was seen causing speciation in the population of finches that resides there. Due to forces of disruptive selection, intermediate beak sizes have been selected against for generations. The resulting population has almost no medium sized beaks. Beak size is important for more than just gathering food, and it has been found that beak size also changes the mating calls of the various finches. Researchers have found that the populations of birds, once a single population, have genetically diverged and are on the tipping point of being considered separate species.

Disruptive Selection in Plants

Famous biologist John Maynard Smith proposed disruptive selection as a method for plant speciation in the late 1960s. The idea is simple and have been applied to many examples since. Many plant traits, such as the color of pea pods, are controlled by individual genes. In a scenario where disruptive selection is affecting a population of plants, the most intermediate individuals are often the heterozygous individual, or those that contain different types of alleles for a gene. Homozygous individuals, on the other hand, have two of the same alleles for a trait. Whether the allele is functional or not, two of the same will produce a phenotype on the extreme end of the spectrum. These individuals will be protected during the disruptive selection, and reproduce more. Over time, the organisms may differ so much that they become reproductively isolated. Often the intermediates were serving the function of transferring genes between the two populations. Without them, in the presence of disruptive selection, speciation can occur.

References

- Dennett, Daniel C. (1991). Consciousness Explained (1st ed.). Boston, MA: Little, Brown and Company. ISBN 978-0-316-18065-8. LCCN 91015614. OCLC 23648691
- Negative-selection-1136: nature.com, Retrieved 03, June 2020
- What-are-some-examples-of-directional-selection: socratic.org, Retrieved 17, July 2020
- Microevolution-and-macroevolution-microevolution: infoplease.com, Retrieved 05, May 2020
- Macroevolution-examples-from-the-primate-world-96679683: nature.com, Retrieved 11, April 2020
- Begon, Michael; Townsend, Colin R.; Harper, John L. (1996). Ecology: Individuals, Populations and Communities (3rd ed.). Oxford; Cambridge, MA: Blackwell Science. ISBN 978-0-632-03801-5. LCCN 95024627. OCLC 32893848
- Genetic-selection-definition-pros-cons: study.com, Retrieved 26, April 2020
- Natural-selection-genetic-drift-and-gene-flow-15186648: nature.com, Retrieved 09, May 2020

Chapter 6

Mutation: An Integrated Study

A change that occurs in any nucleotide of the genome of an organism either due to mistakes during replication or due to environmental factors is known as a mutation. This chapter discusses in detail the various aspects of mutations like mutationism, neutral mutations, point mutation, frameshift mutation, suppressor mutation, dynamic mutation and polar mutation.

Mutation

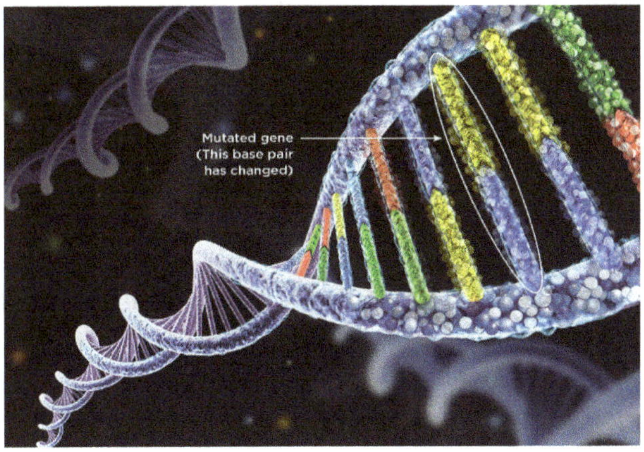

Mutation is an alteration in the genetic material (the genome) of a cell of a living organism or of a virus that is more or less permanent and that can be transmitted to the cell's or the virus's descendants. (The genomes of organisms are all composed of DNA, whereas viral genomes can be of DNA or RNA; Mutation in the DNA of a body cell of a multicellular organism (somatic mutation) may be transmitted to descendant cells by DNA replication and hence result in a sector or patch of cells having abnormal function, an example being cancer. Mutations in egg or sperm cells (germinal mutations) may result in an individual offspring all of whose cells carry the mutation, which often confers some serious malfunction, as in the case of a human genetic disease such as cystic fibrosis. Mutations result either from accidents during the normal chemical transactions of DNA, often during replication, or from exposure to high-energy electromagnetic radiation (e.g., ultraviolet light or X-rays) or particle radiation or to highly reactive chemicals in the environment. Because mutations are random changes, they are expected to be mostly deleterious, but some may be beneficial in certain environments.

In general, mutation is the main source of genetic variation, which is the raw material for evolution by natural selection.

The genome is composed of one to several long molecules of DNA, and mutation can occur potentially anywhere on these molecules at any time. The most serious changes take place in the functional units of DNA, the genes. A mutated form of a gene is called a mutant allele. A gene is typically composed of a regulatory region, which is responsible for turning the gene's transcription on and off at the appropriate times during development, and a coding region, which carries the genetic code for the structure of a functional molecule, generally a protein. A protein is a chain of usually several hundred amino acids. Cells make 20 common amino acids, and it is the unique number and sequence of these that give a protein its specific function. Each amino acid is encoded by a unique sequence, or codon, of three of the four possible base pairs in the DNA (A–T, T–A, G–C, and C–G, the individual letters referring to the four nitrogenous bases adenine, thymine, guanine, and cytosine). Hence, a mutation that changes DNA sequence can change amino acid sequence and in this way potentially reduce or inactivate a protein's function. A change in the DNA sequence of a gene's regulatory region can adversely affect the timing and availability of the gene's protein and also lead to serious cellular malfunction. On the other hand, many mutations are silent, showing no obvious effect at the functional level. Some silent mutations are in the DNA between genes, or they are of a type that results in no significant amino acid changes.

Mutations are of several types. Changes within genes are called point mutations. The simplest kinds are changes to single base pairs, called base-pair substitutions. Many of these substitute an incorrect amino acid in the corresponding position in the encoded protein, and of these a large proportion result in altered protein function. Some base-pair substitutions produce a stop codon. Normally, when a stop codon occurs at the end of a gene, it stops protein synthesis, but, when it occurs in an abnormal position, it can result in a truncated and nonfunctional protein. Another type of simple change, the deletion or insertion of single base pairs, generally has a profound effect on the protein because the protein's synthesis, which is carried out by the reading of triplet codons in a linear fashion from one end of the gene to the other, is thrown off. This change leads to a frame shift in reading the gene such that all amino acids are incorrect from the mutation onward. More-complex combinations of base substitutions, insertions, and deletions can also be observed in some mutant genes.

Mutations that span more than one gene are called chromosomal mutations because they affect the structure, function, and inheritance of whole DNA molecules (microscopically visible in a coiled state as chromosomes). Often these chromosome mutations result from one or more coincident breaks in the DNA molecules of the genome (possibly from exposure to energetic radiation), followed in some cases by faulty rejoining. Some outcomes are large-scale deletions, duplications, inversions, and translocations. In a diploid species (a species, such as human beings, that has a double set of chromosomes in the nucleus of each cell), deletions and duplications alter gene balance and often result in abnormality. Inversions and translocations involve no loss or gain and are functionally normal unless a break occurs within a gene. However, at meiosis (the specialized nuclear divisions that take place during the production of gametes—i.e., eggs and sperm), faulty pairing of an inverted or translocated chromosome set with a normal set can result in gametes and hence progeny with duplications and deletions.

Loss or gain of whole chromosomes results in a condition called aneuploidy. One familiar result of aneuploidy is Down syndrome, a chromosomal disorder in which humans are born with an extra chromosome 21 (and hence bear three copies of that chromosome instead of the usual two). Another type of chromosome mutation is the gain or loss of whole chromosome sets. Gain of sets results in polyploidy—that is, the presence of three, four, or more chromosome sets instead of the usual two. Polyploidy has been a significant force in the evolution of new species of plants and animals.

Most genomes contain mobile DNA elements that move from one location to another. The movement of these elements can cause mutation, either because the element arrives in some crucial location, such as within a gene, or because it promotes large-scale chromosome mutations via recombination between pairs of mobile elements in different locations.

At the level of whole populations of organisms, mutation can be viewed as a constantly dripping faucet introducing mutant alleles into the population, a concept described as mutational pressure. The rate of mutation differs for different genes and organisms. In RNA viruses, such as the human immunodeficiency virus (HIV), replication of the genome takes place within the host cell using a mechanism that is prone to error. Hence, mutation rates in such viruses are high. In general, however, the fate of individual mutant alleles is never certain. Most are eliminated by chance. In some cases a mutant allele can increase in frequency by chance, and then individuals expressing the allele can be subject to selection, either positive or negative. Hence, for any one gene the frequency of a mutant allele in a population is determined by a combination of mutational pressure, selection, and chance.

Mutagens

Chemical Mutagens change the sequence of bases in a DNA gene in a number of ways:

- Mimic the correct nucleotide bases in a DNA molecule, but fail to base pair correctly during DNA replication.
- Remove parts of the nucleotide (such as the amino group on adenine), again causing improper base pairing during DNA replication.
- Add hydrocarbon groups to various nucleotides, also causing incorrect base pairing during DNA replication.

Radiation, High energy radiation from a radioactive material or from X-rays is absorbed by the atoms in water molecules surrounding the DNA. This energy is transferred to the electrons which then fly away from the atom. Left behind is a free radical, which is a highly dangerous and highly reactive molecule that attacks the DNA molecule and alters it in many ways. Radiation can also cause double strand breaks in the DNA molecule, which the cell's repair mechanisms cannot put right.

Sunlight contains ultraviolet radiation (the component that causes a suntan) which, when absorbed by the DNA causes a cross link to form between certain adjacent bases. In most normal cases the cells can repair this damage, but unrepaired dimers of this sort cause the replicating system to skip over the mistake leaving a gap, which is supposed to be filled in later.

Spontaneous mutations occur without exposure to any obvious mutagenic agent. Sometimes DNA nucleotides shift without warning to a different chemical form (know as an isomer) which in turn will form a different series of hydrogen bonds with it's partner. This leads to mistakes at the time of DNA replication.

Causes

Four classes of mutations are:

(1) Spontaneous mutations (molecular decay),

(2) Mutations due to error-prone replication bypass of naturally occurring DNA damage (also called error-prone translesion synthesis),

(3) Errors introduced during DNA repair,

(4) Induced mutations caused by mutagens.

Scientists may also deliberately introduce mutant sequences through DNA manipulation for the sake of scientific experimentation.

One 2017 study claimed that 66% of cancer-causing mutations are random, 29% are due to the environment (the studied population spanned 69 countries), and 5% are inherited.

Humans on average pass 60 new mutations to their children but fathers pass more mutations depending on their age with every year adding two new mutations to a child.

Spontaneous Mutation

Spontaneous mutations occur with non-zero probability even given a healthy, uncontaminated cell. They can be characterized by the specific change:

- Tautomerism: A base is changed by the repositioning of a hydrogen atom, altering the hydrogen bonding pattern of that base, resulting in incorrect base pairing during replication.

- Depurination: Loss of a purine base (A or G) to form an apurinic site (AP site).
- Deamination: Hydrolysis changes a normal base to an atypical base containing a keto group in place of the original amine group. Examples include C → U and A → HX (hypoxanthine), which can be corrected by DNA repair mechanisms; and 5MeC (5-methylcytosine) → T, which is less likely to be detected as a mutation because thymine is a normal DNA base.
- Slipped strand mispairing: Denaturation of the new strand from the template during replication, followed by renaturation in a different spot ("slipping"). This can lead to insertions or deletions.
- Replication slippage.

Error-prone Replication by Pass

There is increasing evidence that the majority of spontaneously arising mutations are due to error-prone replication (translesion synthesis) past DNA damage in the template strand. Naturally occurring oxidative DNA damages arise at least 10,000 times per cell per day in humans and 50,000 times or more per cell per day in rats. In mice, the majority of mutations are caused by translesion synthesis. Likewise, in yeast, Kunz found that more than 60% of the spontaneous single base pair substitutions and deletions were caused by translation synthesis.

Errors Introduced during DNA repair

Although naturally occurring double-strand breaks occur at a relatively low frequency in DNA, their repair often causes mutation. Non-homologous end joining(NHEJ) is a major pathway for repairing double-strand breaks. NHEJ involves removal of a few nucleotides to allow somewhat inaccurate alignment of the two ends for rejoining followed by addition of nucleotides to fill in gaps. As a consequence, NHEJ often introduces mutations.

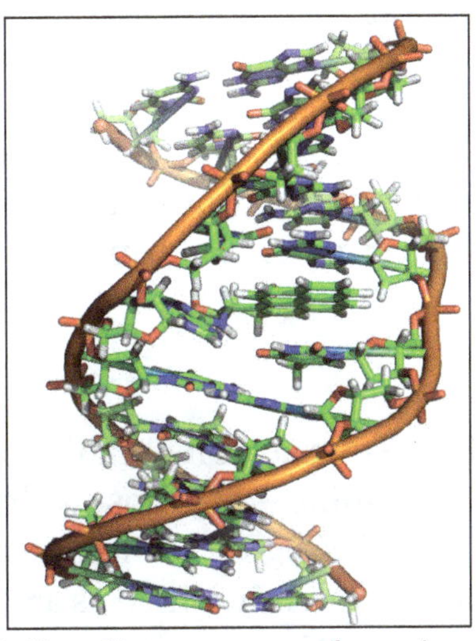

A covalent adduct between the metabolite of benzo pyrene, the major mutagen in tobacco smoke, and DNA.

Induced Mutation

Induced mutations are alterations in the gene after it has come in contact with mutagens and environmental causes.

Induced mutations on the molecular level can be caused by:

- Chemicals
    - Hydroxylamine;
    - Base analogs (e.g., Bromodeoxyuridine (BrdU));
    - Alkylating agents (e.g., N-ethyl-N-nitrosourea (ENU)). These agents can mutate both replicating and non-replicating DNA. In contrast, a base analog can mutate the DNA only when the analog is incorporated in replicating the DNA. Each of these classes of chemical mutagens has certain effects that then lead to transitions, transversions, or deletions.
    - Agents that form DNA adducts (e.g., ochratoxin A);
    - DNA intercalating agents (e.g., ethidium bromide);
    - DNA cross linkers;

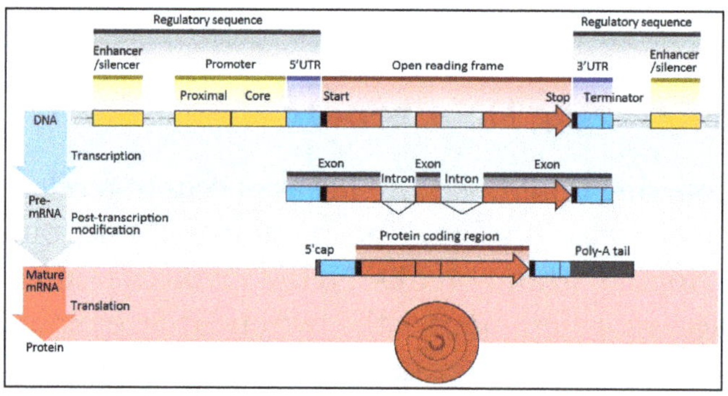

This figure depicts the following processes of transcription, splicing, and translation of a eukaryotic gene.

    - Oxidative Damage;
    - Nitrous acid converts amine groups on A and C to diazo groups, altering their hydrogen bonding patterns, which leads to incorrect base pairing during replication.
- Radiation
    - Ultraviolet light (UV) (non-ionizing radiation). Two nucleotide bases in DNA—cytosine and thymine—are most vulnerable to radiation that can change their properties. UV light can induce adjacent pyrimidine bases in a DNA strand to become covalently joined as a pyrimidine dimer. UV radiation, in particular longer-wave UVA, can also cause oxidative damage to DNA.

- Ionizing radiation. Exposure to ionizing radiation, such as gamma radiation, can result in mutation, possibly resulting in cancer or death.

Classification of Types

By Effect on Structure

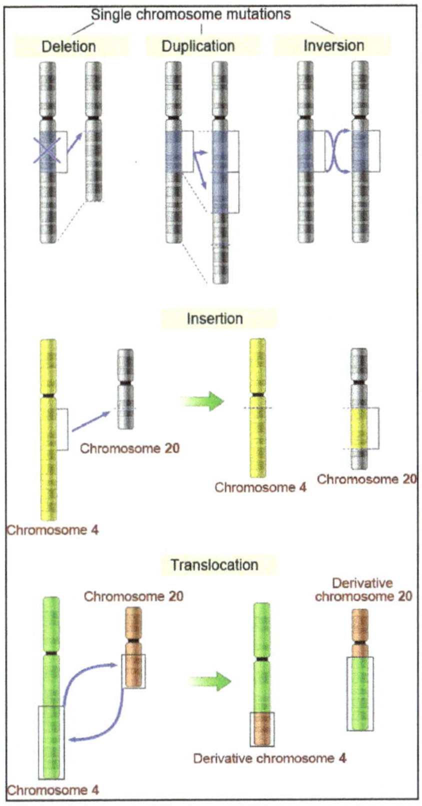

Five types of chromosomal mutations.

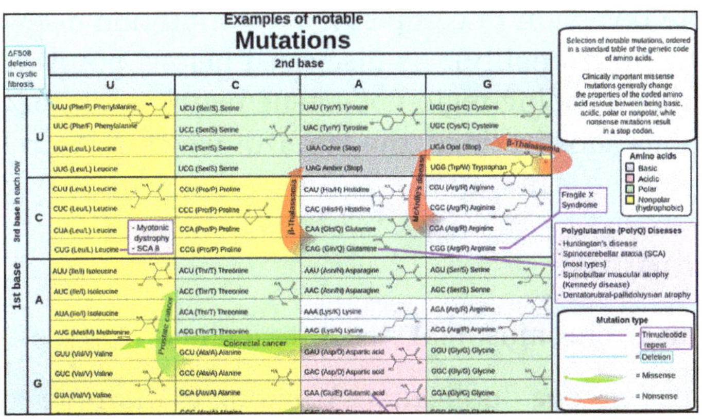

Selection of disease-causing mutations, in a standard table of the genetic code of amino acids.

The sequence of a gene can be altered in a number of ways. Gene mutations have varying effects on health depending on where they occur and whether they alter the function of essential proteins. Mutations in the structure of genes can be classified into several types.

Small-scale Mutations

Small-scale mutations affect a gene in one or a few nucleotides. (If only a single nucleotide is affected, they are called point mutations.) Small-scale mutations include:

- Insertions add one or more extra nucleotides into the DNA. They are usually caused by transposable elements, or errors during replication of repeating elements. Insertions in the coding region of a gene may alter splicing of the mRNA (splice site mutation), or cause a shift in the reading frame (frameshift), both of which can significantly alter the gene product. Insertions can be reversed by excision of the transposable element.
- Deletions remove one or more nucleotides from the DNA. Like insertions, these mutations can alter the reading frame of the gene. In general, they are irreversible: Though exactly the same sequence might in theory be restored by an insertion, transposable elements able to revert a very short deletion (say 1–2 bases) in *any* location either are highly unlikely to exist or do not exist at all.
- Substitution mutations, often caused by chemicals or malfunction of DNA replication, exchange a single nucleotide for another. These changes are classified as transitions or transversions. Most common is the transition that exchanges a purine for a purine (A ↔ G) or a pyrimidine for a pyrimidine, (C ↔ T). A transition can be caused by nitrous acid, base mispairing, or mutagenic base analogs such as BrdU. Less common is a transversion, which exchanges a purine for a pyrimidine or a pyrimidine for a purine (C/T ↔ A/G). An example of a transversion is the conversion of adenine (A) into a cytosine (C). A point mutation are modifications of single base pairs of DNA or other small base pairs within a gene. A point mutation can be reversed by another point mutation, in which the nucleotide is changed back to its original state (true reversion) or by second-site reversion (a complementary mutation elsewhere that results in regained gene functionality). As discussed below, point mutations that occur within the protein coding region of a gene may be classified as synonymous or nonsynonymous substitutions, the latter of which in turn can be divided into missense or nonsense mutations.

Large-scale Mutations

Large-scale mutations in chromosomal structure include:

- Amplifications (or gene duplications) leading to multiple copies of all chromosomal regions, increasing the dosage of the genes located within them.
- Deletions of large chromosomal regions, leading to loss of the genes within those regions.
- Mutations whose effect is to juxtapose previously separate pieces of DNA,

potentially bringing together separate genes to form functionally distinct fusion genes (e.g., bcr-abl).

- Large scale changes to the structure of chromosomes called chromosomal rearrangement that can lead to a decrease of fitness but also to speciation in isolated, inbred populations. These include:
    - Chromosomal translocations: interchange of genetic parts from nonhomologous chromosomes.
    - Chromosomal inversions: reversing the orientation of a chromosomal segment.
    - Non-homologous chromosomal crossover.
    - Interstitial deletions: an intra-chromosomal deletion that removes a segment of DNA from a single chromosome, thereby apposing previously distant genes. For example, cells isolated from a human astrocytoma, a type of brain tumor, were found to have a chromosomal deletion removing sequences between the Fused in Glioblastoma (FIG) gene and the receptor tyrosine kinase (ROS), producing a fusion protein (FIG-ROS). The abnormal FIG-ROS fusion protein has constitutively active kinase activity that causes oncogenic transformation (a transformation from normal cells to cancer cells).
- Loss of heterozygosity: loss of one allele, either by a deletion or a genetic recombination event, in an organism that previously had two different alleles.

By Effect on Function

- Loss-of-function mutations, also called inactivating mutations, result in the gene product having less or no function (being partially or wholly inactivated). When the allele has a complete loss of function (null allele), it is often called an amorph or amorphic mutation in the Muller's morphs schema. Phenotypes associated with such mutations are most often recessive. Exceptions are when the organism is haploid, or when the reduced dosage of a normal gene product is not enough for a normal phenotype (this is called haploinsufficiency).
- Gain-of-function mutations, also called activating mutations, change the gene product such that its effect gets stronger (enhanced activation) or even is superseded by a different and abnormal function. When the new allele is created, a heterozygote containing the newly created allele as well as the original will express the new allele; genetically this defines the mutations as dominant phenotypes. Several of Muller's morphs correspond to gain of function, including hypermorph and neomorph. In December 2017, the U.S. government lifted a temporary ban implemented in 2014 that banned federal funding for any new "gain-of-function" experiments that enhance pathogens "such as Avian influenza, SARS and the Middle East Respiratory Syndrome or MERS viruses."

- Dominant negative mutations (also called antimorphic mutations) have an altered gene product that acts antagonistically to the wild-type allele. These mutations usually result in an altered molecular function (often inactive) and are characterized by a dominant or semi-dominant phenotype. In humans, dominant negative mutations have been implicated in cancer (e.g., mutations in genes p53, ATM, CEBPA and PPARgamma). Marfan syndrome is caused by mutations in the *FBN1* gene, located on chromosome 15, which encodes fibrillin-1, a glycoprotein component of the extracellular matrix. Marfan syndrome is also an example of dominant negative mutation and haploinsufficiency.
- Hypomorphs, after Mullerian classification, are characterized by altered gene products that acts with decreased gene expression compared to the wild type allele.
- Neomorphs are characterized by the control of new protein product synthesis.
- Lethal mutations are mutations that lead to the death of the organisms that carry the mutations.
- A back mutation or reversion is a point mutation that restores the original sequence and hence the original phenotype.

By Effect on Fitness

In applied genetics, it is usual to speak of mutations as either harmful or beneficial:

- A harmful, or deleterious, mutation decreases the fitness of the organism.
- A beneficial or advantageous mutation increases the fitness of the organism. Mutations that promote traits that are desirable are also called beneficial. In theoretical population genetics, it is more usual to speak of mutations as deleterious or advantageous than harmful or beneficial.
- A neutral mutation has no harmful or beneficial effect on the organism. Such mutations occur at a steady rate, forming the basis for the molecular clock. In the neutral theory of molecular evolution, neutral mutations provide genetic drift as the basis for most variation at the molecular level.
- A nearly neutral mutation is a mutation that may be slightly deleterious or advantageous, although most nearly neutral mutations are slightly deleterious.

Distribution of Fitness Effects

Attempts have been made to infer the distribution of fitness effects (DFE) using mutagenesis experiments and theoretical models applied to molecular sequence data. DFE, as used to determine the relative abundance of different types of mutations (i.e., strongly deleterious, nearly neutral or advantageous), is relevant to many evolutionary questions, such as the maintenance of genetic variation, the rate of genomic decay, the maintenance of outcrossing reproduction as opposed to inbreeding and the evolution of sex and genetic recombination. In summary, the DFE plays an important role in predicting evolutionary dynamics. A variety of approaches have been used to study the DFE, including theoretical, experimental and analytical methods.

- Mutagenesis experiment: The direct method to investigate the DFE is to induce mutations and then measure the mutational fitness effects, which has already been done in viruses, bacteria, yeast, and *Drosophila*. For example, most studies of the DFE in viruses used site-directed mutagenesis to create point mutations and measure relative fitness of each mutant. In *Escherichia coli*, one study used transposon mutagenesis to directly measure the fitness of a random insertion of a derivative of Tn10. In yeast, a combined mutagenesis and deep sequencing approach has been developed to generate high-quality systematic mutant libraries and measure fitness in high throughput. However, given that many mutations have effects too small to be detected. and that mutagenesis experiments can detect only mutations of moderately large effect; DNA sequence data analysis can provide valuable information about these mutations.

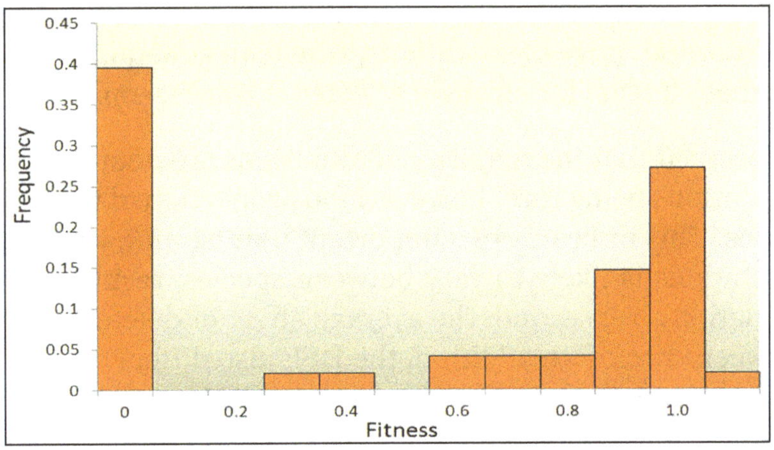

The distribution of fitness effects (DFE) of mutations in vesicular stomatitis virus. In this experiment, random mutations were introduced into the virus by site-directed mutagenesis, and the fitness of each mutant was compared with the ancestral type. A fitness of zero, less than one, one, more than one, respectively, indicates that mutations are lethal, deleterious, neutral, and advantageous.

- Molecular sequence analysis: With rapid development of DNA sequencing technology, an enormous amount of DNA sequence data is available and even more is forthcoming in the future. Various methods have been developed to infer the DFE from DNA sequence data. By examining DNA sequence differences within and between species, we are able to infer various characteristics of the DFE for neutral, deleterious and advantageous mutations. To be specific, the DNA sequence analysis approach allows us to estimate the effects of mutations with very small effects, which are hardly detectable through mutagenesis experiments.

One of the earliest theoretical studies of the distribution of fitness effects was done by Motoo Kimura, an influential theoretical population geneticist. His neutral theory of molecular evolution proposes that most novel mutations will be highly deleterious, with a small fraction being neutral. Hiroshi Akashi more recently proposed a bimodal model for the DFE, with modes centered around highly deleterious and neutral mutations. Both theories agree that the vast majority of novel mutations are neutral or deleterious and that advantageous mutations are rare, which has been supported by experimental results. One example is a study done on the DFE of random mutations in vesicular stomatitis virus. Out of all mutations, 39.6% were lethal, 31.2% were non-lethal deleterious, and 27.1% were neutral. Another example comes from a high throughput mutagenesis experiment with yeast. In this experiment it was shown that the overall DFE is bimodal, with a cluster of neutral mutations, and a broad distribution of deleterious mutations.

Though relatively few mutations are advantageous, those that are play an important role in evolutionary changes. Like neutral mutations, weakly selected advantageous mutations can be lost due to random genetic drift, but strongly selected advantageous mutations are more likely to be fixed. Knowing the DFE of advantageous mutations may lead to increased ability to predict the evolutionary dynamics. Theoretical work on the DFE for advantageous mutations has been done by John H. Gillespie and H. Allen Orr. They proposed that the distribution for advantageous mutations should be exponential under a wide range of conditions, which, in general, has been supported by experimental studies, at least for strongly selected advantageous mutations.

In general, it is accepted that the majority of mutations are neutral or deleterious, with advantageous mutations being rare; however, the proportion of types of mutations varies between species. This indicates two important points: first, the proportion of effectively neutral mutations is likely to vary between species, resulting from dependence on effective population size; second, the average effect of deleterious mutations varies dramatically between species. In addition, the DFE also differs between coding regions and noncoding regions, with the DFE of noncoding DNA containing more weakly selected mutations.

By Impact on Protein Sequence

- A frameshift mutation is a mutation caused by insertion or deletion of a number of nucleotides that is not evenly divisible by three from a DNA sequence. Due to the triplet nature of gene expression by codons, the insertion or deletion can disrupt the reading frame, or the grouping of the codons, resulting in a completely different translation from the original. The earlier in the sequence the deletion or insertion occurs, the more altered the protein produced is. (For example, the code CCU GAC UAC CUA codes for the amino acids proline, aspartic acid, tyrosine, and leucine. If the U in CCU was deleted, the resulting sequence would be CCG ACU ACC UAx, which would instead code for proline, threonine, threonine, and part of another amino acid or perhaps a stop codon (where the x stands for the following nucleotide).) By contrast, any insertion or deletion that is evenly divisible by three is termed an *in-frame mutation*.

- A point substitution mutation results in a change in a single nucleotide and can be either synonymous or nonsynonymous.
    - A synonymous substitution replaces a codon with another codon that codes for the same amino acid, so that the produced amino acid sequence is not modified. Synonymous mutations occur due to the degenerate nature of the genetic code. If this mutation does not result in any phenotypic effects, then it is called silent, but not all synonymous substitutions are silent. (There can also be silent mutations in nucleotides outside of the coding regions, such as the introns, because the exact nucleotide sequence is not as crucial as it is in the coding regions, but these are not considered synonymous substitutions).
    - A nonsynonymous substitution replaces a codon with another codon that codes for a different amino acid, so that the produced amino acid sequence is modified. Nonsynonymous substitutions can in turn be classified as nonsense or missense mutations:
        - ◊ A missense mutation or changes a nucleotide is to cause substitution of a different amino acid. This in turn can render the resulting protein nonfunctional. Such mutations are responsible for diseases such as Epidermolysis bullosa, sickle-cell disease, and SOD1-mediated ALS. On the other hand, if a missense mutation occurs in an amino acid codon that results in the use of a different, but chemically similar, amino acid, then sometimes little or no change is rendered in the protein. For example, a change from AAA to AGA will encode arginine, a chemically similar molecule to the intended lysine. In this latter case the mutation will have little or no effect on phenotype and therefore be neutral.
        - ◊ A nonsense mutation is a point mutation in a sequence of DNA that results in a premature stop codon, or a *nonsense codon* in the transcribed mRNA, and possibly a truncated, and often nonfunctional protein product. This sort of mutation has been linked to different mutations, such as congenital adrenal hyperplasia.

By Inheritance

In multicellular organisms with dedicated reproductive cells, mutations can be subdivided into germline mutations, which can be passed on to descendants through their reproductive cells, and somatic mutations (also called acquired mutations), which involve cells outside the dedicated reproductive group and which are not usually transmitted to descendants.

A mutation has caused this moss rose plant to produce flowers of different colors.
This is a somatic mutation that may also be passed on in the germline.

A germline mutation gives rise to a *constitutional mutation* in the offspring, that is, a mutation that is present in every cell. A constitutional mutation can also occur very soon after fertilization, or continue from a previous constitutional mutation in a parent.

The distinction between germline and somatic mutations is important in animals that have a dedicated germline to produce reproductive cells. However, it is of little value in understanding the effects of mutations in plants, which lack dedicated germline. The distinction is also blurred in those animals that reproduce asexually through mechanisms such as budding, because the cells that give rise to the daughter organisms also give rise to that organism's germline. A new germline mutation not inherited from either parent is called a *de novo* mutation.

Diploid organisms (e.g., humans) contain two copies of each gene—a paternal and a maternal allele. Based on the occurrence of mutation on each chromosome, we may classify mutations into three types:

- A heterozygous mutation is a mutation of only one allele.
- A homozygous mutation is an identical mutation of both the paternal and maternal alleles.
- Compound heterozygous mutations or a genetic compound consists of two different mutations in the paternal and maternal alleles.

A wild type or homozygous non-mutated organism is one in which neither allele is mutated.

Special Classes

- Conditional mutation is a mutation that has wild-type (or less severe) phenotype under certain "permissive" environmental conditions and a mutant phenotype under certain "restrictive" conditions. For example, a temperature-sensitive mutation can cause cell death at high temperature (restrictive condition), but might have no deleterious consequences at a lower temperature (permissive condition). These mutations are non-autonomous, as their manifestation depends upon presence of certain conditions, as opposed to other mutations which appear autonomously. The permissive conditions may be temperature, certain chemicals, light or mutations in other parts of the genome. *In vivo* mechanisms like transcriptional switches can create conditional mutations. For instance, association of Steroid Binding Domain can create a transcriptional switch that can change the expression of a gene based on the presence of a steroid ligand. Conditional mutations have applications in research as they allow control over gene expression. This is especially useful studying diseases in adults by allowing expression after a certain period of growth, thus eliminating the deleterious effect of gene expression seen during stages of development in model organisms. DNA Recombinase systems like Cre-Lox Recombination used in association with promoters that are activated under certain conditions can generate conditional mutations. Dual Recombinase technology can be used to induce multiple conditional mutations to study the diseases which manifest as a result of simultaneous mutations in multiple genes. Certain inteins have been identified which splice only at certain permissive temperatures, leading to improper protein synthesis and thus, loss of function mutations at other temperatures. Conditional mutations may also be used in genetic studies associated with ageing, as the expression can be changed after a certain time period in the organism's lifespan.

- Replication timing quantitative trait loci affects DNA replication.

Nomenclature

In order to categorize a mutation as such, the "normal" sequence must be obtained from the DNA of a "normal" or "healthy" organism (as opposed to a "mutant" or "sick" one), it should be identified and reported; ideally, it should be made publicly available for a straightforward nucleotide-by-nucleotide comparison, and agreed upon by the scientific community or by a group of expert geneticists and biologists, who have the responsibility of establishing the *standard* or so-called "consensus" sequence. This step requires a tremendous scientific effort. Once the consensus sequence is known, the mutations in a genome can be pinpointed, described, and classified. The committee of the Human Genome Variation Society (HGVS) has developed the standard human sequence variant nomenclature, which should be used by researchers and DNA diagnostic centers to generate unambiguous mutation descriptions. In principle, this nomenclature can also be used to describe mutations in other organisms. The nomenclature specifies the type of mutation and base or amino acid changes:

- Nucleotide substitution (e.g., 76A>T) — The number is the position of the nucleotide from the 5' end; the first letter represents the wild-type nucleotide, and the second letter represents the nucleotide that replaced the wild type. In the given example, the adenine at the 76th position was replaced by a thymine.

- If it becomes necessary to differentiate between mutations in genomic DNA, mitochondrial DNA, and RNA, a simple convention is used. For example, if the 100th base of a nucleotide sequence mutated from G to C, then it would be written as g.100G>C if the mutation occurred in genomic DNA, m.100G>C if the mutation occurred in mitochondrial DNA, or r.100g>c if the mutation occurred in RNA. Note that, for mutations in RNA, the nucleotide code is written in lower case.
- Amino acid substitution (e.g., D111E) — The first letter is the one letter code of the wild-type amino acid, the number is the position of the amino acid from the N-terminus, and the second letter is the one letter code of the amino acid present in the mutation. Nonsense mutations are represented with an X for the second amino acid (e.g. D111X).
- Amino acid deletion (e.g., ΔF508) — The Greek letter Δ (delta) indicates a deletion. The letter refers to the amino acid present in the wild type and the number is the position from the N terminus of the amino acid were it to be present as in the wild type.

Prion Mutations

Prions are proteins and do not contain genetic material. However, prion replication has been shown to be subject to mutation and natural selection just like other forms of replication. The human gene PRNP codes for the major prion protein, PrP, and is subject to mutations that can give rise to disease-causing prions.

Somatic Mutations

A change in the genetic structure that is not inherited from a parent, and also not passed to offspring, is called a somatic mutation. *Somatic mutations are not inherited because they do not affect the germline. These types of mutations are usually prompted by environmental causes, such as ultraviolet radiation or any exposure to certain harmful chemicals, and can cause diseases including cancer.*

With plants, some somatic mutations can be propagated without the need for seed production, for example, by grafting and stem cuttings. These type of mutation have led to new types of fruits, such as the “Delicious” apple and the “Washington” navel orange.

Human and mouse somatic cells have a mutation rate more than ten times higher than the germline mutation rate for both species; mice have a higher rate of both somatic and germline mutations per cell division than humans. The disparity in mutation rate between the germline and somatic tissues likely reflects the greater importance of genome maintenance in the germline than in the soma.

Amorphic Mutations

An amorph, a term utilized by Muller in 1932, is a mutated allele, which has lost the ability of the parent (whether wild type or any other type) allele to encode any functional protein. An amorphic mutation may be caused by the replacement of an amino acid that deactivates an enzyme or by the deletion of part of a gene that produces the enzyme.

Cells with heterozygous mutations (one good copy of gene and one mutated copy) may function normally with the unmutated copy until the good copy has been spontaneously somatically mutated. This kind of mutation happens all the time in living organisms, but it is difficult to measure the rate. Measuring this rate is important in predicting the rate at which people may develop cancer.

Point mutations may arise from spontaneous mutations that occur during DNA replication. The rate of mutation may be increased by mutagens. Mutagens can be physical, such as radiation from UV rays, X-rays or extreme heat, or chemical (molecules that misplace base pairs or disrupt the helical shape of DNA). Mutagens associated with cancers are often studied to learn about cancer and its prevention.

Hypomorphic and Hypermorphic Mutations

A hypomorphic mutation is a replacement of amino acids that would hinder enzyme activity, which would reduce the enzyme level but not to the point of complete loss. Usually, hypomorphic mutations are recessive, but haploinsufficiency causes some alleles to be dominant.

A hypermorphic mutation changes the regulation of the gene and causes it to overproduce the gene produce causing a greater than normal enzyme levels. These type of alleles are dominant gain of function type of alleles.

Mutationism

Mutationism refers to historical and contemporary views of evolution that emphasize the role of mutation and that are understood as alternatives to Darwinism. In the mutationist view, change may occur in discrete jumps (i.e., gradualism is not assumed), mutation is seen as the source of novelty (while selection is not seen as creative), and the direction of evolution is understood to reflect both mutation and selection.

By 1909, "mutationism" was referenced as a distinctive rival of Darwinism, following the discovery and naming of "mutation" by Hugo De Vries. At that time, Darwin's mechanism of natural selection was understood to rely on hereditary blending of abundant continuous variations. However, Wilhelm Johannsen's "pure line" experiments appeared to refute this mechanism. Using true-breeding varieties of beans, each with a different size of seeds, Johannsen showed selection could be used to sort out different varieties, but selection within pure lines would not produce evolutionary changes, even though pure lines continued to generate the kind of abundant variations that Darwinians saw as the fuel for evolution.

At the same time, Hugo de Vries's careful studies of wild variants of *Oenothera lamarckiana*, initially published in 1901,showed that distinct new forms could arise suddenly in nature, and could be propagated for many generations without dissipation or blending. De Vries used the term "mutation" to refer to these dramatic variants, as well as any smaller variations that arise via a sudden event and are inherited stably (as distinct from unstable quantitative fluctuations). Thereafter, "mutationism" came to refer to a view of evolution emphasizing that inheritance is discrete, and that hereditary variants originate in discrete events of mutation, which have a role in shaping the outcome of evolution. The terms "saltationism" and "Mendelism" appear to be used similarly. This was contrasted with a Darwinian view in which variation is continuous, and hereditary variations emerge as masses of infinitesimal effects that are shaped by selection.

These and other early discoveries of genetics are often framed relative to a controversy between, on the one hand, early geneticists— the "Mendelians"— including William Bateson, Wilhelm Johannsen, Hugo de Vries, Thomas Hunt Morgan, and Reginald, who advocated Mendelism and mutation, and were understood as opponents of Darwin's original view, and the biometricians and their allies, who opposed Mendelism and were more faithful to Darwin's original vision. The dispute was resolved, and the eclipse of Darwinism ended, with the rise of the "Modern Synthesis" or modern neo-Darwinism.

Neodarwinist View

While historians agree on the above generalities, they disagree on the nature of the critique of Darwinism offered by geneticists, their alternative view of evolution, and the Darwinian restoration in the modern evolutionary synthesis. In the classical view, early geneticists, upon the discovery of mutation, mistakenly assume that all mutations are large, exaggerate their importance, reject Darwin's idea of selection, and imagine an entire theory of evolution based on dramatic mutations, in which new species were created in a single step; this view gains popularity and temporarily eclipses Darwin's theory; it takes decades for scientists to come to their senses and see that genetics and selection may be combined in a way that allows gradual evolution consistent with Darwin's original vision. The classical view, which emerged in Modern Synthesis writings, presents the Modern Synthesis as the received view and mutationism as an obvious error; the decades-long delay in synthesizing genetics and Darwinism is seen as an "inexplicable embarrassment".

The nature of this synthesis was that, due to recombination in the diverse gene pool of a population, selection can shift the phenotype well beyond its initial limits, essentially creating a new form without any mutations. This argument for "the effectiveness of selection" is the centerpiece of Province's seminal history of theoretical population genetics. Thus, in the classical view, scientists who look to distinctive individual mutations for the origin of novelty, rather than to selection smoothly shifting an entire distribution (of many infinitesimal variations) to fit external conditions, are seen as advocates of a failed theory. Mutationism (saltationism), along with Lamarckism and orthogenesis, are presented as anti-Darwinian "blind alleys" separate from the main line leading from Darwin to the present.

Historical View

Though the classical view is commonly repeated by scientists, it is not clear how strongly historians would defend it against revisionist claims. Gayon's Darwinism's Struggle for Survival argues that Darwin's conception of natural selection was incompatible with Mendelian genetics, and that early geneticists are responsible for "the most important event in the history of Darwinism: the Mendelian reconstruction of the principle of selection" as a force that shifts frequencies of discretely inherited types. That is, Gayon places the early geneticists, not in a blind alley, but in the middle of the genetic reformation of evolutionary thinking, validating their main complaints against Darwin's non-Mendelian view of a creative force of selection.

Provine's seminal history of theoretical population genetics presents a version of the classical narrative, yet in the 2001 reprinting of the same work, Provine rejects the inevitability of the Modern Synthesis (which "came unraveled"), and questions both the sufficiency and the logical coherence of its selection-and-recombination-driven "gene pool" view of population genetics. Provine lists a variety of post-Synthesis findings that are not consistent with the synthetic view of population genetics presented in his 1971 analysis.

The most sympathetic view of the mutationists is given by Stoltzfus and Cable, who argue that early geneticists accepted selection, allowed gradual evolution, and laid the conceptual foundations for the genetical view of evolution that prevails today. They argue that the fusion of Darwinism and genetics that Provine describes achieved premature acceptance, but later proved untenable and is no longer the foundation of contemporary evolutionary thinking, which is better aligned with the "Mendelian-mutationism" of early geneticists.

In summary, classical historiography holds that mutationism is a failed view in which the role of mutation in evolution is misperceived, while the role of selection is wrongly dismissed. In revisionist views, the mutationists accept both mutation and selection, and assign them the same qualitative roles they are assigned today, though perhaps not the same quantitative emphasis. At the time of the Darwin centennial in Cambridge in 1909, Mutationism and Lamarckism were contrasted with Darwin's "Natural Selection" as competing ideas; 50 years later, at the University of Chicago centennial of the publication of *On the Origin of Species*, mutationism (like Lamarckism) was no longer seriously considered. Nevertheless, after another 50 years, evolutionary biologists are reconsidering the mutationist view.

Neutral Mutation

Neutral mutation is a mutation that has no selective advantage or disadvantage. Considerable controversy surrounds the question of whether such mutations can exist.

When the idea of a constant molecular clock first emerged, it was thought that the predominant evolutionary force underlying amino acid or nucleotide substitutions was natural selection. Following this line of thinking, a constant molecular clock would indicate that adaptive substitutions in different species occur constantly over time. However, it is hard to explain how adaptive substitutions would occur in such a clock-like manner. Theoretically, the fates of adaptive mutations are determined by several evolutionary parameters, such as the strength of the selective advantage of that mutation, the size of the effective population, and adaptive mutation rates. These parameters are likely to differ between species, and even within a species, depending upon specific mutations and their interactions with environments.

Instead, Kimura proposed that most changes at the molecular level have little functional consequences, or are 'neutral'. If a mutation has no fitness consequence, its fate in the population is determined completely by random chance. This means that we cannot predict whether a specific neutral mutation will eventually be fixed in the population. However, the *rate* at which neutral substitutions occur in the population can be predicted because it depends upon a single parameter, namely the mutation rate.

Let's imagine a population with N number of haploid individuals. If neutral mutations occur at rate u per individual per generation, the total number of mutations in one generation will be N times u. Since all these new mutations are neutral, their fates are completely determined by chance. In other words, all mutations have equal chance of reaching fixation (which leads to a 'substitution'). The probability that each new neutral mutation will reach fixation, given that a substitution occurred, is simply $1/N$. The rate of substitutions is calculated as the number of new mutations in each generation (Nu) multiplied by the probability each new mutation reaches fixation ($1/N$), which equals u. In other words, for neutral mutations, the rate of substitution is equal to the rate of mutation.

Almost all the controversies at the heart of debates over neutral molecular clocks stem from the question of what the major sources of mutations are. This question is directly relevant to understanding patterns of mutation, which are the ultimate source of evolutionary change and genetic disease. Furthermore, understanding how mutation rates vary between lineages and within genomes is a fundamental question in comparative genomics, which aims to use sequence comparisons to identify genomic regions that are functionally important.

So what determines neutral mutation rates? One of the most important contributors to neutral molecular clocks is lineage-specific variation in generation times. From early on, the idea of a constant neutral molecular clock was perceived as being at odds with the molecular mechanisms of germline mutation. It has long been considered that most mutations arise from errors in DNA replication in gremlines. Since mutations occur when germline DNA is replicated for the next generation, they should accumulate in proportion to the number of generations, rather than the absolute amount of time. Therefore, if we compared the numbers of substitutions that have accumulated in two lineages since their divergence, the lineage with longer generation time, having undergone fewer DNA replication events, would harbor fewer substitutions compared to the lineage with the shorter generation time. Consequently, the molecular clock should run more slowly in species with longer generation times. This idea is referred to as the 'generation-time effect'.

Point Mutation

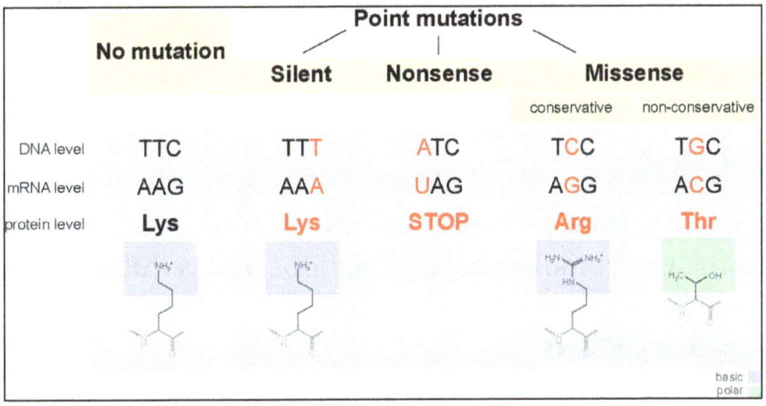

Point mutation is a mutation in which change within a gene in which one base pair in the DNA sequence is altered. Point mutations are frequently the result of mistakes made during DNA replication, although modification of DNA, such as through exposure to X-rays or to ultraviolet radiation, also can induce point mutations.

There are two types of point mutations: transition mutations and transversion mutations. Transition mutations occur when a pyrimidine base (i.e., thymine [T] or cytosine [C]) substitutes for another pyrimidine base or when a purine base (i.e., adenine [A] or guanine [G]) substitutes for another purine base. In double-stranded DNA each of the bases pairs with a specific partner on the corresponding strand—A pairs with T and C pairs with G. Thus, an example of a transition mutation is a GC base pair that replaces a wild type (or naturally occurring) AT base pair. In contrast, transversion mutations occur when a purine base substitutes for a pyrimidine base, or vice versa; for example, when a TA or CG pair replaces the wild type AT pair.

At the level of translation, when RNA copied from DNA is converted into a string of acids during protein synthesis, point mutations often manifest as functional changes in the final protein product. Thus, there exist functional groupings for point mutations. These groupings are divided into silent mutations, missense mutations, and nonsense mutations.

Missense Mutation

A missense mutation is when the change of a single base pair causes the substitution of a different amino acid in the resulting protein. This amino acid substitution may have no effect, or it may render the protein nonfunctional.

The necessity to understand the effects of missense mutations becomes clear when one considers genetic diseases. It is known that the substitution of only one residue in a protein sequence can be related to a number of pathological conditions such as Alzheimer's, Parkinson's and Creutzfeldt-Jakob's diseases.[6] It is also well known that accumulation of autosomal mutations can lead to cancers, and that hereditary diseases are caused by one or more germline mutations. The analysis of the impacts of missense mutations advances our understanding of the relationships between protein structure and function, and allows us to decipher the mechanisms of the effects of disease mutations and thereby pathogenesis.With personalized medicine on the not-so-distant horizon, it is swiftly becoming essential to understand the process by which genetic mutations lead to disease.

The plausible effects of missense mutations range from affecting the macromolecular stability to perturbing macromolecular interactions and cellular localization.

Point nonsense Mutation

A nonsense mutation is the substitution of a single base pair that leads to the appearance of a stop codon where previously there was a codon specifying an amino acid. The presence of this premature stop codon results in the production of a shortened, and likely nonfunctional, protein.

Frameshift Mutation

A frameshift mutation is a genetic mutation caused by a deletion or insertion in a DNA sequence that shifts the way the sequence is read. A DNA sequence is a chain of many smaller molecules called nucleotides. DNA (or RNA) nucleotide sequences are read three nucleotides at a time in units called codons, and each codon corresponds to a specific amino acid or stop signal. During translation, the sequence of codons is read in order from the nucleotide sequence to synthesize a chain of amino acids and form a protein. Frameshift mutations arise when the normal sequence of codons is disrupted by the insertion or deletion of one or more nucleotides, provided that the number of nucleotides added or removed is not a multiple of three. For instance, if just one nucleotide is deleted from the sequence, then all of the codons including and after the mutation will have a disrupted reading frame. This can result in the incorporation of many incorrect amino acids into the protein. In contrast, if three nucleotides are inserted or deleted, there will be no shift in the codon reading frame; however, there will be either one extra or one missing amino acid in the final protein. Therefore, frameshift mutations result in abnormal protein products with an incorrect amino acid sequence that can be either longer or shorter than the normal protein.

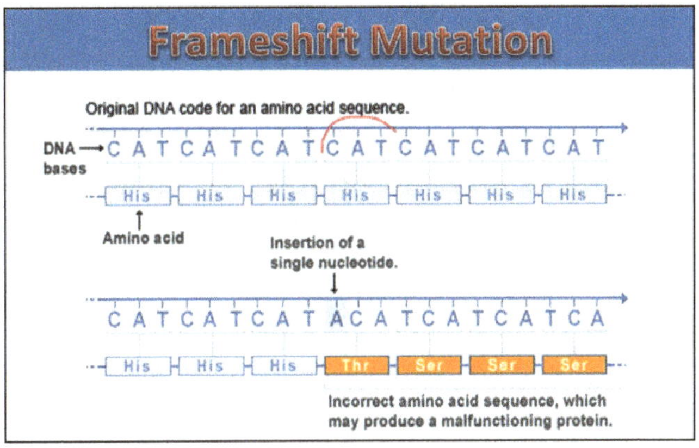

Suppressor Mutation

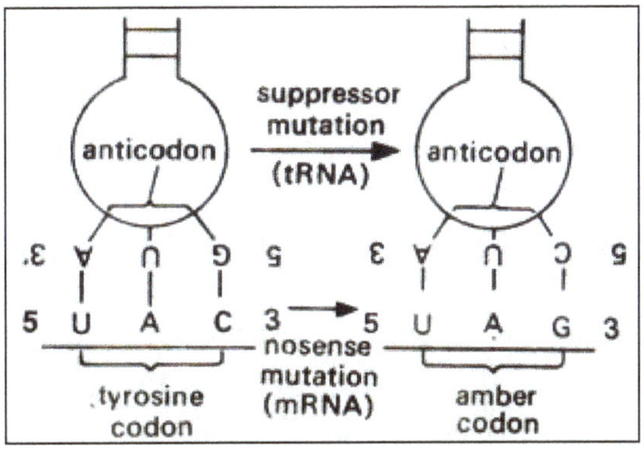

When the effect of one mutation is suppressed by another mutation, the latter is called a suppressor mutation. For instance, AABB may be wild, and aaBB a mutant; then if aabb, is also a wild type, the mutation in 'B' will be called suppression mutation, because the effect of 'a' is suppressed by 'b' Mechanism of this suppression effect was not known for some time, but now its mechanism is largely known. These effects can be conveniently classified into:

1. Intragenic suppression,
2. Intergenic suppression.

Intragenic Suppression

Suppression of a mutation at one site within a gene can be achieved by mutation at another site within the same gene. For instance, a mutant A46 that produces inactive tryptophan syntheses enzyme, due to substitution of glutamic acid for glycine, at position 210 may be corrected or suppressed due to mutation A446 involving substitution of cysteine in place of tyrosine at position 174. These two mutants A46 and A446 independently give rise to inactive enzyme, but if present together give rise to active enzyme, thus suppressing the mutant effect of each other. In many other combinations, double mutants do not show any suppression effect. The suppression effect really depends on whether or not the second alteration, can restore the 3-D configuration of the enzyme. There are a number of known examples of such intragenic or intracistronic suppression or complementation. The reader is advised to consult an advanced book for further details in this connection.

Intergenic suppression, which means suppression of mutation in one gene due to mutation in another gene, can be achieved mainly due to alteration in transfer RNA (tRNA). The suppression may also be effective by alteration brought about in ribosome conformation.

There are atleast four types of such suppression mutations, which are discussed below-

Nonsense suppression and mutant tRNA. Sometimes due to a change in anticodon of a specific tRNA, it may develop the capacity to read the mutant terminating codon (amber or UAG, ochre or UAA and opal or UGA mutants). Since these nonsense mutations cause premature termination of polypeptide chain, they can be suppressed if a mutant tRNA can read the terminating codon and substitute an amino acid at the location of termination codon. Mutant tRNAs are known which are capable of recognizing termination codons and insert amino acids like tyrosine, serine or glutamine to restore the function of protein. However these mutant tRNAs may also read the normal stop signals of proteins thus causing elongation of protein beyond its normal termination thus causing disturbances. To prevent this disturbance, double stop signals are often available in mRNA, so that if mutant tRNA reads one, the other may still bring about termination. Moreover, two genes may code for same tRNA, so that if one (minor gene) mutates, and gives rise to suppressor tRNA the other (major gene) still functions and gives rise to normal tRNA.

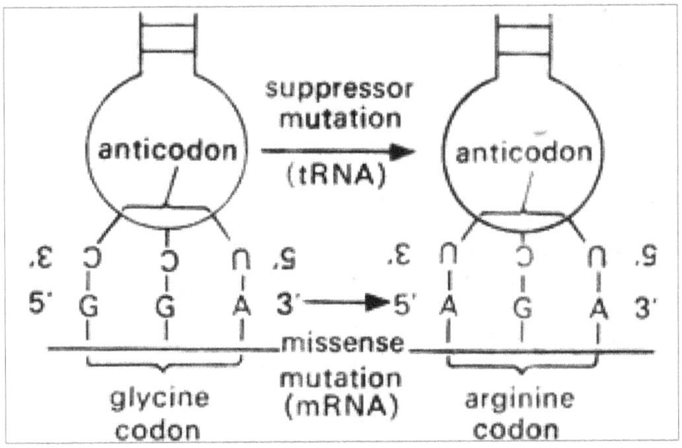

Figure: Suppressor mutation involving a change in one base of anticodon of tyrosine tRNA leading to reading of termination amber codon UAG, which resulted from nonsense mutation in tyrosine codon UAC thus leading to suppression.

Transfer RNA (tRNA) mediated UAG suppression was first observed, when in coat protein of phage R 17, amber codon 5'UAG3' could be read by a tyrosine suppresser tRNA, in which the anticodon had mutated from 5'GUA3' to 5'CUA3'. There are also other examples of this type of suppression.

Missense Suppression and Mutant tRNA

Missense Mutations

Will lead to synthesis of a protein with a wrong amino acid at a particular site. Such a mutation can also be suppressed like nonsense mutations, by mutant tRNA, which will misread the codon and thus substitute a correct amino acid, although the mutant codon codes for a different amino acid. For instance, in mutant A36 in tryptophan synthetase of E. coli, glycine (coded by GGA) is replaced by arginine (coded by AGA). A suppresser mutant may have tRNA, depositing glycine for codon AGA actually meant for arginine.

Frameshift Mutations and Mutant tRNA

It has been shown that a frameshift mutation due to a deletion can be suppressed by mutant tRNA having an anticodon of two bases rather than three (doublet anticodon rather than triplet). Similarly, addition frameshift mutations can be suppressed by mutant tRNAs each having a quadruplet anticodon instead of a triplet. Mutant tRNAs with doublet and quadruplet anticodons have actually been detected in biological systems.

Suppression and Ribosomal Mutations

Mutations in some ribosomal proteins may cause a change in ribosome structure, so that it may now misread the mRNA codons. One such mutation, called *ram* (ribosomal ambiguity) is capable of reading all the three termination codons, thus causing suppression of nonsense mutations. Some frameshift mutations may also be suppressed by these *ram* mutants.

Dynamic Mutation

The term 'dynamic mutation' was introduced to distinguish the unique properties of expanding, unstable DNA repeat sequences from other forms of mutation. The past decade has seen dynamic mutations uncovered as the molecular basis for a growing number of human genetic diseases and for all of the characterized 'rare' chromosomal fragile sites. The common properties of the repeats in different diseases and fragile sites have given insight into this unique form of DNA instability. While the dynamic mutation mechanism explains some unusual genetic characteristics, unexpected findings have raised new questions and challenged some assumptions about the pathways that lead from mutation to disease.

The process of dynamic mutation is affected by a variety of elements and factors that have been divided into those directly associated with the expanding repeat (cis-acting elements) and those (trans-acting factors) whose interaction with the repeat contributes to its instability. Examples of cis-acting factors include the copy number and the composition of the repeat (whether it is perfect or interrupted). In general, higher copy number alleles that are free of interruptions (perfect repeats) are more unstable than those of lower copy number and contain interruptions (imperfect repeats). While there is indirect evidence for the existence of trans-acting factors (such as parental gender bias in repeat instability), the identity of these factors remains elusive. Candidates include the proteins involved in DNA replication such as FEN1 (Rad27), which is known to have a role in Okazaki fragment metabolism.

Dynamic Mutation Undergoing Repeat Sequences

The first expanded repeats to be identified were the trinucleotides CCG/CGG and CAG/CTG. Initially this was taken as evidence that only trinucleotide repeats could undergo this form of mutation (which was sometimes referred to as trinucleotide repeat expansion). Furthermore, since these two repeats could form secondary structures, it was assumed that this was also a necessary condition for repeat instability. Subsequently, six of the ten possible trinucleotide repeats have been found to exhibit repeat expansion in humans, and, more recently, 5 and 6 bp microsatellite repeats and 12, 33 and 42 bp minisatellite repeats have also been observed to undergo repeat copy number expansion.

Founder effects are observed in association with dynamic mutation as a higher frequency of particular alleles in the affected population, compared with the unaffected population, suggesting that these alleles are 'at-risk' for mutation. The likelihood is that the initial mutation on these alleles is a rare event that increases the instability of the repeat, setting in process the subsequent expansion of these few founder mutations. Founder effects are a common feature of dynamic mutation and in many cases common flanking marker haplotypes provide evidence of a relationship between the expanded alleles and the longest normal (usually uninterrupted) alleles in the population. Therefore a common molecular mechanism appears to be responsible.

At the *FRAXA* locus, detailed haplotyping studies, together with sequence analysis of different repeat alleles, have provided evidence in support of the view that certain alleles are predisposed to instability. Similar analyses in other populations suggest that such differences are likely to be population-specific rather than indicative of a molecular pathway predisposed by *cis*-acting elements in and around the *FRAXA* repeat. One possible explanation for the inconsistency here is that chance has a role to play and that the differences between populations merely reflect the different rare initial events that have precipitated the dynamic mutation process. Given that there are multiple possibilities for generating an unstable repeat then different populations are likely to have different combinations and/or frequencies of these unstable alleles. Thus, in terms of understanding the mutation process, caution is needed when extrapolating the results observed in one population to the human species as a whole.

Factors Contributing to Repetition in DNA Expansion

Gender bias in the instability of repeats during their transmission from parent to offspring is a common property of dynamic mutations. Most cases of the congenital form of myotonic dystrophy are maternally inherited, while the juvenile onset Huntington disease is generally from paternal inheritance. The gender of the germ line clearly contributes to these biases; however, the identity of the factors involved remains obscure.

Other factors that interact in some way with the repeat are also thought to contribute to repeat instability. A variety of circumstantial evidence (particularly from studies in model systems) suggests that Okazaki fragments may have a role to play in the expansion of DNA repeats. Several of the repeats (e.g. CCG in *FRAXA* and CAG in *DM*) exhibit a bimodal distribution of instability with the boundary between small changes and large changes in copy number approximating Okazaki fragment length. In *Escherichia coli* (which has a substantially longer Okazaki fragment length), this boundary is increased in copy number again to that approximating Okazaki fragment length. The secondary structure of CAG/CTG and CGG repeat sequences has been found to inhibit Okazaki flap endonuclease (FEN1) binding and cleavage. In addition, the FEN1 homologue, Rad27, has a role to play in DNA repeat instability in yeast. Yeast Rad27 mutants are prone to both DNA breakage and copy number instability of repeat sequences.

Contradicting the proposed role of DNA replication, Kennedy and Shelbourne found dramatic expansion of the CAG repeat in the striatum of a transgenic mouse HD model. These neurons are presumably no longer undergoing cell division, indicating that expansion is occurring independently of DNA replication. Similarly, Kotvun and McMurray have found that germ-line trinucleotide repeat sequence expansion in transgenic mice also proceeded in the absence of DNA replication via gap repair. Perhaps, given that DNA instability is occurring at the repeat sequence in the absence of replication, some form of transcriptional healing may be contributing to the repeat expansion process. Transcriptional healing has been invoked in the fragile site expression of tandemly repeated small RNA transcripts. In further support of a role for breakage and repair Manley have found that *Msh2* is required for CAG repeat somatic cell instability in an HD transgenic mouse.

Age-dependent somatic cell instability of the CTG repeat sequence has been found in myotonic dystrophy, although here it is not confined to those cells that are affected in the disease and in fact gives a growth advantage to lymphoblastoid cell lines that have undergone expansion.

Pathogenic Pathways

Loss of Function/Haploinsufficiency

Extinction of, or interference with transcription is one means by which expanded repeats cause loss of gene function. Examples are: fragile X syndrome, in which repeat expansion brings about methylation of the *FMR1* promoter region; myoclonus epilepsy (*EPM1*), in which repeat expansion physically separates transcription factors in the cystatin B promoter; and Friedreich ataxia, in which the intronic expanded repeat sequence adopts a 'sticky' DNA structure that interferes with transcription of the frataxin gene. While these resultant diseases exhibit recessive (or X-linked) transmission, the potential exists for dominant transmission where reduction in the level or activity of a rate limiting gene product can appear as haploinsufficiency. This appears to be the case for the *SIX5* gene in myotonic dystrophy, as the expanded CTG repeat is located in the *SIX5* promoter. The reduction in effective 'dosage' of the *SIX5* gene product brought about by extinction of one allele is thought to contribute to some aspects of pathology.

Gain of Function: The Toxic Polyglutamine Hypothesis—Old Concerns about a New Dogma

Numerous expanded repeat sequences are located within coding regions where they translate into expanded polyglutamine tracts in disease alleles. This, together with a substantial body of data demonstrating the toxicity of polyglutamine in cells, has led to the view that the expanded polyglutamine is a common, crucial component in the pathogenesis of these diseases. In support of this view, an antibody, 1C2, raised against the normal length polyglutamine tract of the TATA-box binding protein (TBP), was found to specifically recognize the expanded polyglutamine in several of the expanded repeat diseases, e.g. huntingtin, ataxin 1 and others. This was taken as evidence that the expanded polyglutamine adopted a different conformation that in some way contributed to its toxicity. The context in which the repeat sequence was located was proposed to account for why such an otherwise toxic conformation did not cause problems when located within TBP. This theory is now somewhat flawed, as the expanded polyglutamine tract in some alleles (n >50) of TBP has recently been associated with neurodegenerative disease. It is possible that yet another conformation of the expanded polyglutamine tract is adopted once the repeat sequence gets beyond an even greater threshold.

Nuclear inclusions were found in the affected neurons of HD and SCA1 transgenic mouse models and HD patients, prompting the assertion that these expanded polyglutamine-containing nuclear inclusions cause neuronal dysfunction. Klemment deleted the SCA1 self-association region in transgenic mice and found that these mice still exhibited ataxia and Purkinje cell pathology; however, no evidence of nuclear aggregates was found. Other studies on the brains of individuals affected with Huntington's disease revealed that nuclear aggregates were more commonly found in unaffected neurons than in the vulnerable striatal spiny neurons. Consequently, it has now been proposed that rather than being pathogenic, nuclear inclusions may even be protective against the toxic effects of the expanded polyglutamine.

Further conflicting evidence to a common role for polyglutamine in pathogenesis is that disease alleles of the CAG repeat in spinocerebellar ataxia 6 are well within the normal range for other 'polyglutamine diseases'. Functional assays on expanded CAG alleles of the SCA6 associated gene, *CACNA1A*, demonstrate increased Ca^{2+} transport and thus make it likely that this gain of function is a specific and unique cause of pathogenesis for this disorder. Taken together, these exceptions suggest that the notion of a common single pathogenic pathway involving polyglutamine is unlikely.

As if contradictions with polyglutamine encoding genes are not enough, a growing list of repeat-associated ataxias do not even encode polyglutamine. In *SCA12*,the expanded CAG repeat is located within the 5′-untranslated region (5′-UTR) of the PPP2R2B gene, while in *SCA8* an expanded CUG is located within the 3′-UTR of a transcript (although there is controversy over whether this expansion is causative of neurodegenerative disease). In addition, an intronic expanded 5 bp repeat was recently found to be associated with *SCA10*.

So the question must be asked whether expanded polyglutamine is a necessary and sufficient condition for those diseases in which it has been identified as the disease-causing agent. If it is, then how is the cellular specificity of this toxicity accounted for when at least some of these proteins are widely expressed? One possible explanation for this conundrum comes from the findings of Kennedy and Shelbourne, which reveal affected cell-specific amplification of the unstable CAG repeat in a mouse model of Huntington's disease. Individuals affected with Huntington's disease have also been found to exhibit somatic instability, which was most pronounced in the affected areas of the brain. This suggests that ongoing somatic mutation may contribute to diseases caused by expanded repeats. This expansion may also account for the relationship between germ-line repeat copy number and age-at-onset for these diseases. The time taken for the disease to manifest could represent how long it takes for the disease-associated allele to reach a critical higher copy number threshold in the affected cell population.

What then is the evidence that polyglutamine has a role? The principal data are from transgenic animal model studies, particularly those from *SCA1* transgenic mice. Mutation of the ataxin-1 nuclear localization signal severely diminishes toxicity, whereas abolition of sequences necessary for aggregation does not. Crossing of *SCA1* mice with a transgenic mouse deficient in ubiquitin ligase results in more severe pathology. For *SCA2* and *SCA6*, nuclear localization does not appear to be necessary, suggesting different pathways than that for SCA1 (and HD).

In an HD transgenic mouse model where transcription of the transgene could be artificially turned on or off, expression of the expanded CAG-containing gene was elegantly demonstrated to be necessary for HD-like pathology. This animal model provides the first indication that HD (and hopefully other expanded CAG repeat diseases) is potentially treatable—given that, in those patients at risk, the disease-causing gene can either be switched off or the toxic gene product can be cleared from the sensitive neurons.

Presumably other factors, such as modification of the polyglutamine tract and the protein context in which the polyglutamine is located, contribute to the cellular specificity of toxicity. Examples of where this apparently is or is not the case have been reported. The relative contribution of various intrinsic and extrinsic factors is therefore likely to differ between the different disease genes (and experimental model systems). Genetic screens for factors that modify polyglutamine-generated pathology in *Drosophila* have produced a large number and variety of candidates, and it will be of great interest to see which (if any) of these has a role in human polyglutamine disease.

Bias of ascertainment was invoked by Penrose to account for (and discredit) the observation of anticipation seen by clinical geneticists in myotonic dystrophy families. While expanded DNA repeats now give a molecular basis (and validation) for the phenomenon of anticipation, the relative ease with which expanded CAG repeats can be screened for as a cause of neurodegenerative disease may well have led to an overemphasis of the relative importance of this form of mutation and perhaps contributed to the view that expanded CAG repeats necessarily have a common pathogenic pathway (via their encoded expanded polyglutamines). While there is a substantial body of evidence in support of a role for polyglutamine in some repeat expansion diseases there is a growing list of exceptions and inconsistencies which suggest (at the least) that this is not necessarily a common pathogenic pathway exhibited by all neurodegenerative diseases in which an expanded repeat (let alone an expanded CAG) is the disease-causing mutation.

Further Evidence for Alternative Pathways—Multiple Gene Effects

In myotonic dystrophy there is good evidence to support not only a role for multiple genes, but also multiple pathways in mediating the pathogenic effects of repeat expansion. Three genes (*DMPK, SIX5* and *DMWD*) are all located in the immediate vicinity of the DM expanded CTG repeat. At least three distinct pathogenic pathways are thought to contribute to DM.

1. The expanded repeat affects the splicing of the DMPK gene in which it is located.
2. The RNA transcript containing the expanded repeat is both inappropriately compartmentalized and titrates a crucial protein (muscle blind). Expression of the CTG repeat in another muscle-specific transcript is able to cause at least some of the DM phenotype (myotonia and myopathy).
3. The DM CTG repeat is also located in the promoter region of the *SIX5* gene and its expansion also interferes with the expression of this gene. The *SIX5* gene encodes a transcription factor whose concentration is critical for eye, muscle and testicular development. Heterozygous deletion of *Six5* is sufficient to cause ocular cataracts in transgenic mice, and therefore the *SIX5* haploinsufficiency brought about by the DM CTG repeat expansion is likely to contribute to the DM phenotype.

The *FRAXE* chromosomal fragile site is associated with a mild form of mental retardation. The expanded CCG repeat was first initially located in the 5′-UTR of a gene, referred to as *FMR2*. The expression of *FMR2* is extinguished by repeat expansion and subsequent methylation of the CpG island promoter region, in a manner analogous to the silencing of the *FMR1* gene at the *FRAXA* locus in fragile X syndrome. It was therefore thought that the loss of *FMR2* was responsible for the phenotype associated with *FRAXE*. However, an additional gene, *FMR3*, has now been identified that shares the same methylated CpG island. Expression of *FMR3* is also silenced by *FRAXE* full mutation and therefore this gene may also contribute to the phenotype.

Polar Mutation

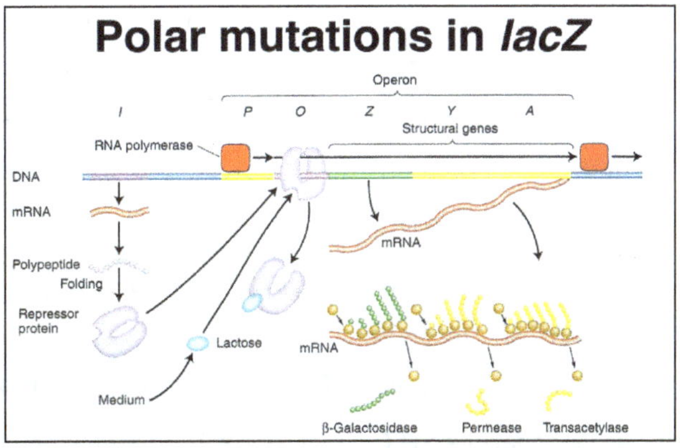

Polar mutation is a mutation that affects the transcription or translation of part of the gene or operon downstream of the mutant site.

The first class of polar mutations involves an "insertion" of a foreign piece of DNA which contains an RNA termination signal. Such a mutation eliminates downstream gene expression since no RNA polymerase will read through the "new" termination signal to reach those genes. One therefore typically sees a complete cessation of downstream mRNA synthesis. Gene expression typically refers to the act of transcription of the gene in question.

The second class of polar mutations are those whose primary effect is the premature termination of protein synthesis. This often leads indirectly to a partial reduction of downstream mRNA synthesis. Such polarity requires the presence of a functional Rho protein, which is normally involved in transcription termination at the end of operons. A model explaining the mechanism by which premature protein stop signals, termed nonsense mutations, Rho factor causes RNA polymerase to terminate transcription downstream from a site of translation termination if the following conditions exist:

i. A region of the mRNA, lacking ribosomes because of the upstream nonsense signal, contains a site that activates the ATP synthase activity of Rho, probably by binding Rho. There is *in vitro* evidence that such sites are C- rich and G-poor (in the mRNA) and that an approximately 12-base spacing of C residues may be important, though a precise consensus is unclear. This latter number is intriguing since biochemical characterization suggests that Rho complexes with about 13 nucleotides per protein subunit of the hexamer.

ii. At some distance downstream (deliberately vague, but probably not important) from this site, the activated Rho "catches up" with RNA polymerase (at a polymerase pause site?) and causes it to terminate transcription.

This effect requires that there has not been translation initiation, proper or improper, between the "upstream site" and the termination region, or else ribosomes will mask the sites on the mRNA. It seems that termination sites occur fairly frequently and that the "upstream sites" may be the limiting factors in determining where polarity occurs.

The model is the following: if there is untranslated, unstructured mRNA, then there is a probability that Rho will bind there (the probability will be bigger as the "open" mRNA gets longer and when it has sequences that are preferred by Rho). Having bound, there is a probability that it will catch up with the transcribing RNAP and cause termination

References

- Orengo CA, Thornton JM (July 2005). «Protein families and their evolution-a structural perspective». Annual Review of Biochemistry. 74: 867–900. doi:10.1146/annurev.biochem.74.082803.133029. PMID 15954844
- Mutation-genetics, science: britannica.com, Retrieved 25, March 2020
- Carroll SB, Grenier JK, Weatherbee SD (2005). From DNA to Diversity: Molecular Genetics and the Evolution of Animal Design (2nd ed.). Malden, MA: Blackwell Publishing. ISBN 978-1-4051-1950-4. LCCN 2003027991. OCLC 53972564

- Point-mutation, science: britannica.com, Retrieved 06, May 2020
- Bohidar HB (January 2015). Fundamentals of Polymer Physics and Molecular Biophysics. Cambridge University Press. ISBN 978-1-316-09302-3
- Neutrality-and-molecular-clocks-100492542: nature.com, Retrieved 13, June 2020
- Long M, Betrán E, Thornton K, Wang W (November 2003). "The origin of new genes: glimpses from the young and old". Nature Reviews Genetics. 4 (11): 865–75. doi:10.1038/nrg1204. PMID 14634634
- Polar-mutation, molecular-biology-glossary-10974: genscript.com, Retrieved 17, July 2020
- Frameshift-mutation-frame-shift-mutation-frameshift-203: nature.com, Retrieved 19, May 2020

Permissions

All chapters in this book are published with permission under the Creative Commons Attribution Share Alike License or equivalent. Every chapter published in this book has been scrutinized by our experts. Their significance has been extensively debated. The topics covered herein carry significant information for a comprehensive understanding. They may even be implemented as practical applications or may be referred to as a beginning point for further studies.

We would like to thank the editorial team for lending their expertise to make the book truly unique. They have played a crucial role in the development of this book. Without their invaluable contributions this book wouldn't have been possible. They have made vital efforts to compile up to date information on the varied aspects of this subject to make this book a valuable addition to the collection of many professionals and students.

This book was conceptualized with the vision of imparting up-to-date and integrated information in this field. To ensure the same, a matchless editorial board was set up. Every individual on the board went through rigorous rounds of assessment to prove their worth. After which they invested a large part of their time researching and compiling the most relevant data for our readers.

The editorial board has been involved in producing this book since its inception. They have spent rigorous hours researching and exploring the diverse topics which have resulted in the successful publishing of this book. They have passed on their knowledge of decades through this book. To expedite this challenging task, the publisher supported the team at every step. A small team of assistant editors was also appointed to further simplify the editing procedure and attain best results for the readers.

Apart from the editorial board, the designing team has also invested a significant amount of their time in understanding the subject and creating the most relevant covers. They scrutinized every image to scout for the most suitable representation of the subject and create an appropriate cover for the book.

The publishing team has been an ardent support to the editorial, designing and production team. Their endless efforts to recruit the best for this project, has resulted in the accomplishment of this book. They are a veteran in the field of academics and their pool of knowledge is as vast as their experience in printing. Their expertise and guidance has proved useful at every step. Their uncompromising quality standards have made this book an exceptional effort. Their encouragement from time to time has been an inspiration for everyone.

The publisher and the editorial board hope that this book will prove to be a valuable piece of knowledge for students, practitioners and scholars across the globe.

Index

www.ingramcontent.com/pod-product-compliance
Lightning Source LLC
LaVergne TN
LVHW080906210926
848485LV00004B/601